# 鼎爺廚房

## 家傳粵式手工菜

# 作者簡介

李家鼎（鼎爺），香港著名藝員，自幼受父親培訓，操練出一身好廚藝；同時由於天賦、興趣和不屈不撓的熱血精神，武藝與馬術亦相當精湛，可謂廚武雙全。

於 2016 年冬，他主持烹飪節目《阿爺廚房》，在節目中鼎爺大顯身手，親自下廚，炮製拿手的粵式小菜、湯水及甜品。他的刀工、他的廚藝，讓觀眾認識、欣賞鼎爺的另一面。

2017 年夏，他出版首本食譜《鼎爺廚房～家傳粵式手工菜》大受歡迎，在半年內再版十多次，成為當年綜合圖書類別銷量冠軍。

2018 年夏，鼎爺推出第二本食譜《鼎爺廚房 2～懷舊風味撚手菜》，繼續發揚粵菜的精巧和美好。

# 前｜言

## 鼎爺的粵式手工菜

李家鼎（鼎爺），相信大部分讀者的腦海中，浮現的第一印象必定是武藝超凡。毋庸置疑，自青年時代已拜師習武、具多年武術及馬術指導經驗、曾參與無數電影及電視劇的動作設計及演出的他，其專業出色的武術形象早就深入民心。

除武藝出色外，鼎爺的廚藝也不遑多讓。

鼎爺出生於廣州，相信對飲食有研究的朋友們都會知道，廣州人對於「飲和食」都是相當講究，當然鼎爺已去世的父親亦不例外！鼎爺父親從事酒樓業，對食材、刀工和配搭都有嚴厲的要求，鼎爺幼承庭訓、耳濡目染下，加上自己的努力，廚藝相當了得。鼎爺的家訓是「食不言，寢不語」，講求餐桌禮儀，不鋒芒過露，故縱使鼎爺身懷精湛廚藝，亦只有至親和好友知道。

去年，鼎爺榮升祖父，亦因為孫兒的誕生，令鼎爺萌生出版食譜書的念頭，讓家傳的粵菜精粹能一代一代傳下去，讓子孫及讀者們認識粵菜的精巧。從鼎爺親自將英文書名改為 Grandpa's Kitchen，中文書名亦改為《家傳粵式手工菜》，可見一斑。這書分為小菜、米麵、湯和糖水，除精采的圖片、詳盡的文字演繹外，還有「鼎爺話你知」的技巧分享，讓讀者們能夠懂得和掌握這些小細節，明白菜式箇中的竅門，烹調時事半功倍。

現在就隨鼎爺一起進入其家傳粵式手工菜的世界吧！

# 目錄 Contents

# 鼎爺廚房的 基本要求

鼎爺烹調菜式以煮出菜鮮香為主，講求用料新鮮，調味以簡單為原則。他廚房內只備有基本的調味料和香料，例如：豉油、老抽、生粉、黃砂糖、粗鹽、醋、花椒、八角、桂皮、香葉、胡椒粉、草果、薑、蔥頭、紅頭蔥、蒜頭、豆豉、麵醬、柱侯醬、麻油、五香粉、米酒、冰糖等。

其中鼎爺特別推崇紅頭蔥，皆因它蔥味濃郁。

此外，鼎爺還會自製幼鹽和淮鹽，詳細做法看第 12-13 頁。

鼎爺經常強調：煮餸不能一本通書看到老，因每家爐火、每人口味各不同，所以要了解你家的爐火和菜式定要試味。

以下是鼎爺與你分享的一些入廚心得。

## 鼎爺的好拍檔

### 松木砧板

砧板，鼎爺以松木砧板為首選，皆因它質地較軟，斬硬的食材如豬骨、雞鴨、大魚等，不會彈刀；切肉絲、蔥粒等等時不會跳刀。如砧板的木質過硬，會在斬食材時傷刀。

使用砧板前，他會在檯面先放一塊微濕布，再放上砧板，這樣是防止砧板在使用時「走位」。

砧板如果清潔不當，很容易變為細菌溫床。日常清潔須將醋和粗鹽撒在砧板上，並用刷子擦洗砧板，沖水，再風乾。

## 刀

在拍攝食譜的過程中，鼎爺只使用兩把刀，分別是 2 號桑刀和文武刀。

以前 2 號桑刀是用來切細桑葉給蠶蟲吃，現在則用來切絲、切薄片。文武刀使用廣泛，斬、剁、切皆可，如廚房面積細，建議只備文武刀就可以了。

建議要買有木柄的菜刀，因使用時不容易滑手，兼易用力。

在廚房也需要備有磨刀石，就算多鋒利的刀也會有變鈍的一天。

## 鼎爺的揸刀姿勢

### 正確的揸刀姿勢

手指公按着刀，食指傍着刀的另一面，其餘手指握着刀柄，利用後面的手掌肉推動切東西，每次切完後要放鬆力度，這樣才不容易邊，否則手部會很容易抽筋。

### 切東西時，手指的安全姿勢

切東西時，將中指屈曲頂着刀身，無名指和食指按着食材，一邊切食材、手指一邊後褪，因有中指頂着刀身，這樣就不會切傷手指。初次學習切東西的朋友，宜刀口斜斜地向出，待熟練後才直刀切東西。

## 蒸 的 基 本 技 巧

首次聽見「放死氣」都頗為嚇人，究竟是甚麼東東？

原來凡是肉類、魚類等都會有股「翕」味或腥臭味，這股異味會在蒸時散發出來，積聚在鍋內，故在蒸後3、4分鐘須打開鍋蓋，讓這股死氣散去。然後蓋上鍋蓋繼續蒸，食材的鮮甜味就會「跑出來」。

經驗是累積出來的！

多次看見鼎爺打開蒸鍋，俯身吹走蒸氣，然後才放入食材；這比「放死氣」還令人好奇，原來此舉是避免蒸氣遮眼，看得清楚些。

估不到鼎爺也有頑皮的一面。

## 調味的先後次序

調味的先後次序，是菜式美味與否的關鍵。

凡是將食材上底味或菜式的調味，鼎爺都是由淡味開始，除了有補救的機會外，也避免將食材「喇」死，變得油鹽不進。

鼎爺的調味基本功是：先下糖，將食材軟化；跟着下油，讓食材吸收油分，令質感滑些；再下生粉，讓生粉裹着食材彷似一個保護罩般，讓跟着下的重味料如醋、豉油、鹽等不會直闖進食材內，因為調味的目的是增鮮，而不是破壞原味。

## 湯料為甚麼要飛水？

對於煲湯，凍水下湯料還是大滾水下湯料？鼎爺認為兩者都可以。

但堅持，肉類在煲湯前一定要飛水，而且要在凍水下鍋，水由凍至暖，由暖至熱，由熱至滾，食材內深層的血水和污垢會慢慢逼出來。如在滾水下鍋，只將食材表面的血水和污垢煮出來，但在深層的血水和污垢因肉類的表面鎖緊了，而不能釋放出來，但經長時間煲老火湯時，血水和污垢就會隨熱力走出來，令湯渾濁。

# 鼎爺最愛用明火

明火煮食，是鼎爺的堅持。

鼎爺可能給讀者的印象是有火氣，原來他煮餸更講究鑊氣，指定要用明火，認為用明火煮出來的菜式才有鑊氣。而有鑊氣的首要條件，是「爐要旺、火要猛」，因為唯有火力十足的明火爐頭，方可將整個炒鑊均勻加熱，在炒菜時，即使是「拋鑊」，鑊底也能接觸到明火，不致炒鑊降溫，影響外觀和味道。因爐旺火猛，炒菜的過程快，佳餚自然有「鑊氣」。

用明火另一個優點是「看得見、易控制」，能隨心所欲地調節火力，不用「靠估」，無論是蒸、煎、炒、炸、燜、煲湯等等皆宜。以我們常用的烹調法為例，蒸海鮮、蒸雞或肉質較堅韌的食材，宜用猛火蒸；雞蛋、蔬菜及瓜果等鮮嫩易熟的食材，宜用中火慢蒸。煎魚宜以細火慢煎，使其外皮焦脆，內裏鮮嫩；如煎雞翼、牛扒等肉類先用大火煎以鎖緊表面水分，轉而以中火煎至中央漸熟。炸的烹調法可分三個階段，首階段用滾油大火猛炸，目的是要令外層入口香脆，之後轉以中細火浸炸，務求令食材中央部分熟透，離鑊前再用大火猛炸，迫出食材內多餘油分，保持酥脆口感之餘，亦不會滿口油花，較為健康；當然，要爐火的火力猛才能做出好效果。

至於廣東人最愛的老火湯，更是無「火」不行。控制火候是煲湯的主要關鍵，唯有借助明火，方能均勻地加熱整個湯鍋，拉近鍋底與鍋側的溫差，令沸騰的對流現象四方八面地同時進行，令食材的味道完全溶入湯水內。

鼎爺為你介紹的菜式，皆使用「看得見、易控制」的明火，就算是初入廚者，也能從中取得成功感，領略烹調的樂趣。

幼鹽
TABLE SALT

淮鹽
SPICED SALT

## 幼 鹽

鼎爺日常用的幼鹽，都會經
過加工才採用。
他將粗鹽用白鑊炒至起粉，倒
入臼內，再春成幼鹽狀，用粉
篩篩去粗粒即可用。

## 淮 鹽

用來蘸芝麻蝦餅、酥炸生蠔
等的淮鹽，鼎爺也是自製的。
做法如加工幼鹽般，將粗鹽
用白鑊炒至起粉，倒入臼內，
再春成幼鹽狀，用粉篩篩去
粗粒，舀入五香粉，拌勻即
成淮鹽。

# 自家製馬拉盞

## HOMEMADE BELACAN SAUCE

將馬拉盞切塊，用燒魚網半烘半燒馬拉盞（去除羶臭味），
使馬拉盞鬆身就可以。放馬拉盞入碗內，攤涼，加入油拌勻
成糊狀。

## 材　料

| | |
|---|---|
| 馬拉盞 | 80 克 |
| 葱頭 | 6 粒（切碎） |
| 蒜頭 | 2 個（剁碎） |
| 蝦米 | 半斤（浸腍後切碎） |
| 瑤柱 | 12 粒（浸軟後撕碎） |
| 銀魚仔 | 6 兩（不要沖洗） |
| 糖 | 3 湯匙 |
| 紅辣椒 | 3 隻（切碎） |

## 做 法

1. 用滾油炸銀魚仔，炸至魚仔開始變色，撈起，攤涼後略剁碎。圖 1~3

2. 處理蝦米、瑤柱等材料。圖 4~5

3. 熱鑊下油，倒入一半蒜蓉和葱頭蓉，下蝦米和瑤柱，炒至蝦米呈金黃色，下銀魚仔和糖，炒至差不多乾身時先下少許馬拉盞，炒勻後再下其餘的馬拉盞，倒入剩下的蒜蓉和葱頭蓉，炒至將近收乾時下紅辣椒粒，熄火，在鑊內攤平馬拉盞，用餘溫將馬拉盞烘一烘，味道會更香。圖 6~10

### 鼎 爺 話 你 知

◆ 用網燒馬拉盞，可以去除羶臭味外，也
  可以使馬拉盞鬆身，能與油調成糊狀，
  炒時不會結成一團。燒馬拉盞時要用微
  細火，否則會很易燒焦。

◆ 如沒有燒魚網，可以用平底鑊慢慢烘馬
  拉盞。

◆ 宜往售賣東南亞食材的店舖購買一磚磚
  的馬拉盞，切勿用蝦膏頂替。

# 梅子蒸蟹
STEAMED CRAB WITH PICKLED PLUMS

蟹最肥美是農曆正月、五月、九月，
這次用的是紅花蟹，肉質飽滿，味道鮮甜，宜清蒸。

| 材 | 料 | | 梅 | 子 | 醬 |
|---|---|---|---|---|---|

| 材料 | |
|---|---|
| 紅花蟹 | 2隻（約5吋） |
| 中國檸檬 | 1個（吃時才榨汁） |

| 梅子醬 | |
|---|---|
| 梅子 | 2粒 |
| 糖 | 1茶匙 |
| 蒜蓉 | 1茶匙 |
| 薑蓉 | 半茶匙 |
| 米酒 | 少許 |
| 生粉 | 少許 |
| 油 | 1茶匙 |
| 鹽 | 少許 |

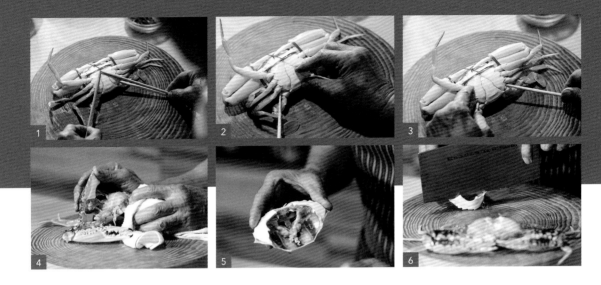

## 做 | 法

1. 梅子醬做法：梅子去核，剁成蓉，下糖、蒜蓉、薑蓉與梅子蓉拌勻，加少許酒，調勻；下生粉和少許油（將味道包圍，若太早落，其他的不能入味），最後下鹽。試味，如太酸，再下多點糖。

2. 劏蟹：要在蟹的最尾的兩隻腳之間，插入兩枝尖筷子，到心臟即死 圖1~3。先揭開蟹蓋，用手指推出蟹胃，並棄去沙囊，削去蟹蓋邊（如不削去，會花較長時間將蟹蒸熟）。去鰓，切去蟹箝，剝去蟹臍，因不是黃油蟹可切去少許蟹爪尖；用刀背拍裂蟹箝（凸的一面向砧板）。圖4~10

3. 蒸蟹時蟹膏向上，放梅子蓉在蟹上，蒸10分鐘，不用放死氣；因劏時已清去不好的味道。圖11~12

4. 享用時擠上中國檸檬汁。

7

8

9

10

11

12

## 鼎 爺 話 你 知

◆ 蟹胃是非常寒涼的，一定要棄去。

◆ 蒸蟹時間，視乎蟹的大小而定。

# 蝴蝶過河蒸石斑

STEAMED CROUCHING GAROUPA

## 材料

| | |
|---|---|
| 石斑 | 1 條（約 1.5 斤） |
| 竹筷子 | 1 對 |
| 薑絲 | |
| 葱絲 | 適量 |
| 滾油 | |

## 甜豉油

| | |
|---|---|
| 豉油 | 各適量，煮至糖溶化 |
| 糖 | |

## 做　法

1. 魚劏開（劏至近魚尾），洗淨。圖 1~2

2. 碟上頭尾各放一隻竹筷子，放上魚（掰開魚腹，讓其趴在竹筷子上），尾部蓋上小碟。圖 3~4

3. 用大火蒸魚 9 分鐘（蒸至 3-4 分鐘開蓋放死氣），將魚取出，移去小碟及筷子；放一箸薑絲及葱絲在魚上，慢慢從魚頭開始淋滾油（因為魚頭需要熟一點），甜豉油淋在碟邊。圖 5~8

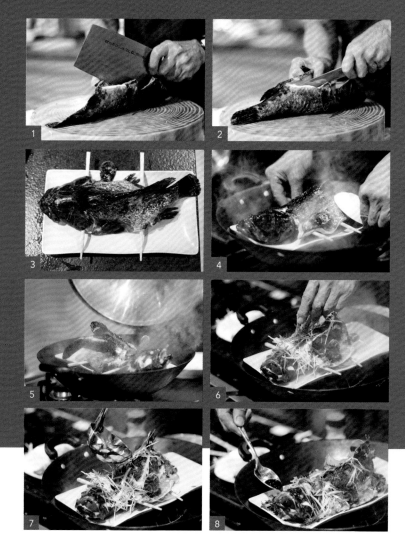

### 鼎 爺 話 你 知

◆ 因魚尾的肉比較薄，放上小碟使魚尾不致直接受熱而過熟。

◆ 將魚切成蝴蝶形再趴在竹筷子上蒸，讓魚能受熱均勻。

◆ 魚蒸熟後不要立即揭蓋，因魚肉開始收縮，如一下子遇冷，魚肉會收縮得更快，當再下滾油時，就不好吃了。

◆ 最優質的豉油也會有點苦澀味，但將豉油煮滾，苦澀味就會消失；至於下多少糖，視乎你的口味而定。

# 酥炸生蠔
## DEEP-FRIED BATTERED OYSTERS

## 材　料

| | |
|---|---|
| 美國桶蠔 | 2 桶 |
| 豬網油 | 1-2 張 |

生粉 / 麵粉  
唥汁　　　　　適量  
淮鹽  
（做法參考 p.13）

## 洗　蠔　料

生粉

冷飯

## 黏　漿

生粉糊

## 炸　漿

| | | |
|---|---|---|
| 麵粉 | 3 湯匙 | |
| 生粉 | 1 湯匙 | 拌勻至 |
| 蛋白 | 2 個 | 沒有粉粒 |
| 水 | 適量 | |

## 做法

1. 蠔用洗蠔料揉去污垢，洗淨，放入滾水內飛水至三成熟，撈起，立刻沖凍水，用毛巾吸乾水分。圖2

2. 豬網油攤開，切開成長方形（長度是能包裹整隻蠔）。

3. 蠔蘸生粉，用豬網油捲着，以生粉糊封口，再蘸生粉，裹上炸漿，放入熱油內（油溫不要太高，否則外熟內生）炸至微黃盛起。圖3~9

4. 用大火翻炸蠔，將蠔內的油迫出，炸至金黃色即撈起，伴准鹽、喼汁享用。

### 鼎爺話你知

◆ 豬網油買回來後，一定要浸水，否則豬網油會溶掉。靚的豬網油以原整一塊、沒有太多破洞、缺口，拿起來不會軟塌塌的為佳。圖1

◆ 這道菜的外層講求酥脆，所以要將蠔飛水迫出內裏水分，避免酥炸生蠔變得軟腍腍。

◆ 要用花生油炸蠔，蠔才會甘香脆口。

◆ 不要一下子放太多蠔進炸油內，否則會令油溫下降，令蠔吸收太多油，吃時會滿口油。

# 蒸蒜蓉豆豉欖角蟠龍鱔

STEAMED COILED WHITE EEL WITH
GARLIC, BLACK BEANS AND OLIVES

宜購買鮮活的白鱔，
同時白鱔宰後要盡快
食用，因放置過久肉
質會起變化，容易中
毒。

## 材料

| | | | |
|---|---|---|---|
| 白鱔 | 1 條（約 2 呎） | 生粉 | 2 茶匙 |
| 荷葉 | 2 張（浸軟） | 油 | 1 湯匙 |
| 豆豉 | 1 湯匙（切粒） | 鹽 | 半茶匙 |
| 欖角 | 半湯匙（切粒） | 陳皮 | 1/4 個 |
| 薑蓉 | 1 茶匙 | 葱粒 | 適量 |
| 蒜蓉 | 3 茶匙 | 芫茜 | 適量 |
| 糖 | 1 茶匙 | | |

## 做 法

1. 請魚販只劏白鱔,不要去潺,回家後才用粗鹽、百潔布去潺,洗淨。

2. 從白鱔的背部落刀,要切斷骨但鱔肚是相連的厚件圖 1~3;用生粉、油抹勻白鱔(在表皮就可,不需抹在肉內),醃約 10 分鐘。

3. 陳皮浸軟後刮去瓤,半份切絲、半份剁蓉;拌勻豆豉、欖角、陳皮、薑蓉、蒜蓉、生粉、油和糖,並下少許鹽成豆豉蒜蓉醬。

4. 切去荷葉的柄部(避免放入白鱔時隆起,放上荷葉遮蓋此洞),在荷葉上刺孔。放上白鱔,並捲成圈狀,塗上豆豉蒜蓉醬,蓋上荷葉。圖 4~8

5. 用大火蒸約 12 分鐘;白鱔蒸熟後不要立刻移離蒸鑊,大約 3 分鐘後才取出,讓鱔肉吸回蒸汁。

6. 取出,澆熟油和豉油,撒下葱粒和芫茜,趁熱享用。圖 9~12

◆ 豆豉宜切粒，不宜舂爛，讓豆豉的味道慢慢滲入鱔肉中。

◆ 豆豉蒜蓉醬內的薑蓉不宜多，夠辟去腥味就可，否則會蓋過蒜味。

◆ 為何陳皮要切絲和剁蓉？因為陳皮蓉在蒸時可滲入鱔肉內，而陳皮絲則在咀嚼時有口感。

◆ 蒸白鱔時用荷葉墊底和鋪面，味道會特別清香。同時不用放死氣，因荷葉在蒸前已刺孔。

# 三絲燜斑尾

BRAISED GAROUPA WITH SHREDDED PORK,
MUSHROOMS AND WOOD EAR FUNGUS

## 材料

| | |
|---|---|
| 石斑尾 | 1 條 |
| 梅頭豬肉 | 2 兩 |
| 木耳 | 1/4 兩 |
| 冬菇 | 2 朵 |
| 榨菜 | 1/4 個 |
| 芹菜 | 1 棵 |
| 紅辣椒 | 1 隻 |
| 薑 | 2 片 |
| 芫茜 | 少許 |

## 汁料

| | |
|---|---|
| 上湯 | 1 杯 |
| 老抽 | 半湯匙 |
| 糖 | 1 茶匙 |
| 鹽 | 少許 |
| 生粉 | 1 茶匙 |

1

2

3

4

5

## 做法

1. 在魚尾的厚肉部位剞幾刀，撲上生粉，泡油至7成熟。圖 1~3
2. 豬肉切絲；木耳浸軟後切絲；冬菇浸軟後切絲；榨菜切絲；撕去芹菜硬筋後，切絲；紅辣椒去籽，切絲；薑片切絲。
3. 豬肉絲用糖、油和生粉拌勻，下鹽拌勻，醃約10 分鐘。
4. 熱鑊下油，倒入豬肉絲炒至 6 成熟，下冬菇絲、榨菜絲，跟着下木耳絲，兜勻，盛起備用。
5. 熱鑊下油，放入薑絲和魚尾，倒入已調勻的汁料，煮滾後放入做法（4）的絲料，燜約 15 分鐘，放上芹菜絲、紅辣椒和芫茜同燜一會即可。圖 4~5

### 鼎爺話你知

◆ 在魚尾的厚肉部位剆幾刀，可以令熱力滲透，魚肉會容易熟。

◆ 如不確定魚尾是否已熟，可用竹籤刺在厚肉部位測試。

# 茄汁煎中蝦

## FRIED MEDIUM PRAWNS IN TOMATO SAUCE

## 材 料

| | |
|---|---|
| 急凍中蝦 | 10隻（約5吋長） |
| 青、紅長椒 | 各1隻（切塊） |
| 洋葱 | 半個（切角） |
| 蒜蓉 | 半粒份量 |
| 乾葱蓉 | 1粒份量 |
| 茄膏 | 1湯匙 |
| 番茄 | 1個（切角） |
| 糖 | |
| 鹽 | 各適量，要試味 |
| 唸汁 | |
| 中國檸檬 | 1個（榨汁） |

## 做 法

1. 剪去蝦槍、蝦尾、頭至眼的部位，去腳。在蝦的第一和第二節之間，剪一下，在最尾一節和上一節之間剪一下，在第三節之間挑去蝦腸，但別太大力，因蝦腸很易斷。挑去蝦腸後，用絞剪在頭下剪至第三節，但別剪至尾部。

2. 用乾布吸去蝦的水分。

3. 蝦走油至不超過5成熟（就算不走油，乾煎也可），撈起。

4. 熱鑊下油，爆炒青紅椒、洋葱，下蒜蓉、乾葱蓉，炒勻，先加入茄膏，下番茄，倒入蝦，加糖（多點也不怕，因現時的味道較酸），拌勻，下少許鹽兜勻，灑入唸汁拌勻，熄火，上碟，趁熱擠上檸檬汁。

---

### 鼎 爺 話 你 知

◆ 剪去蝦槍、蝦尾、頭至眼的部位和去腳，是防止炸蝦時濺油，引起危險。

◆ 用較剪在蝦頭下剪至第三節，除了容易入味外，也較易熟，因這部位的肉質較厚。

---

# 薑葱火腩生蠔

OYSTERS WITH ROAST PORK BELLY,
GINGER AND SPRING ONION IN CLAY POT

小菜　薑葱火腩生蠔

## 材　料

| | |
|---|---|
| 美國桶蠔 | 2桶 |
| 火腩 | 半斤 |
| 薑片 | 3片（拍裂） |
| 紅葱頭 | 1粒 |
| 蒜頭 | 1粒 |
| 葱段 | 1棵 |
| 芫茜 | 適量 |
| 紅椒絲 | |

## 洗　蠔　料

生粉、冷飯

## 調　味　料

| | |
|---|---|
| 糖 | 3茶匙 |
| 豉油 | 2茶匙 |
| 老抽 | 1茶匙（調色，最後下） |
| 玫瑰露酒 | 適量 |

### 鼎　爺　話　你　知

◆ 薑片拍裂後才爆生蠔，好讓薑味釋出。

◆ 用冷飯和生粉洗蠔，蠔會特別乾淨；因藏在蠔裙、蠔隙的污垢會被冷飯帶出來。勿用鹽洗蠔，因蠔煮後會縮水。

◆ 蠔一定要拖水，避免煮時出水。

◆ 烹調生蠔要準確掌握時間，過火會韌。

## 做　法

1. 用生粉、冷飯揉去蠔的污垢。蠔洗淨後拖水，用毛巾或廚房紙吸乾水分。

2. 砂鍋下油，爆香薑片、紅葱頭和蒜頭，加入火腩，待火腩有少許焦邊，下蠔，兜至蠔半熟，先下糖炒勻，再下豉油調味，先下少許老抽調色，如顏色不足才下多點，蓋上砂鍋蓋焗一會。

3. 放入葱段和芫茜，蓋上砂鍋蓋，在鍋蓋灑玫瑰露酒，讓酒香迫入食材內。熄火，趁熱享用。

4. 享用時，撒上紅椒絲裝飾。

# 西蘭花韭黃炒斑球

## STIR-FRIED GAROUPA FILLET WITH BROCCOLI AND YELLOW CHINESE CHIVES

## 材　料

| | |
|---|---|
| 石斑 | 1 條（約斤半） |
| 西蘭花 | 1 個 |
| 韭黃 | 2 兩（切度） |
| 薑片 | 1 小片 |
| 紅椒絲 | 少許 |

## 醃　魚　料

糖
熟油
生粉
鹽

各少許，調味份量視乎
起出多少斑肉，但醃魚
次序要先糖、油、生粉，
最後下鹽。

## 做　法

1. 西蘭花切成一朵朵，用醋、粗鹽和水浸半小時 圖 1~2，撈起，飛水，待水再滾後下少許糖以除去草青苦澀味；西蘭花有 5、6 成熟後，下少許鹽提升鮮味，再下熟油，撈起。

2. 石斑起肉，切塊，剩下的魚骨可用來煲湯 圖 3~4。先用糖、熟油和生粉拌勻魚塊，拌勻後才下鹽。

3. 熱鑊下油，倒入魚塊，炒勻，下薑片，炒至魚約 6、7 成熟，下韭黃（先下較老的部分）炒勻，再下其餘韭黃，兜一兜就可下西蘭花，兜勻，下少許糖，炒勻後下少許鹽兜勻，熄火，上碟，放紅椒絲裝飾。圖 5~9

---

### 鼎 爺 話 你 知

◆ 西蘭花用醋、粗鹽和水浸半小時，有殺菌、消毒和去蟲的效用。

◆ 將西蘭花撈起後立刻沖凍水，西蘭花可保持原色而不會變黃。

◆ 醃魚片時宜用熟油，否則進食時會有油腥味。

◆ 凡有韭黃的菜式皆不宜潷酒，否則韭黃會淡而無味。

# 蒸三及第魚

STEAMED GRASS CARP BELLY WITH
SALTED FISH AND DRIED SHRIMPS

## 材 料

| | |
|---|---|
| 鯇魚腩 | 1 件 |
| 蝦乾 | 1 兩（沖淨，略浸後瀝乾） |
| 梅香鹹魚（中間部分） | 1 件 |
| 薑絲 | ⎤ |
| 糖 | ⎬ 適量 |
| 紅椒（切片） | ⎦ |

## 調 味 汁 料

| | |
|---|---|
| 熟油 | ⎤ 適量 |
| 豉油 | ⎦ |

## 做 法

1. 梅香鹹魚起骨，不要魚肚，用刀壓一壓魚肉，讓鹹魚更加出味。圖 1~2

2. 碟內放入蝦乾，覆上魚腩，鋪上鹹魚，放薑絲、紅椒，下少許糖。

3. 用大火煲滾水，放入魚蒸 13-14 分鐘，蒸約 3 分鐘時放死氣。圖 3

4. 熄火後，不要立刻拿出來，在鑊內 1-2 分鐘讓魚肉入味。

5. 在魚旁下熟油和豉油，趁熱享用。圖 4

### 鼎 爺 話 你 知

◆ 這道菜的魷魚腩採用軟邊部分，魚脂的甘香完全被蝦乾吸收；而魷魚腩則吸收了梅香鹹魚的鹹鮮味，不是賣花讚花香，實在是非常美味。

◆ 在鹹魚面放些糖，可去除鹹魚皮的苦澀味。

◆ 蒸鹹魚一定要下薑絲，但如蒸新鮮魚則不可以下，因為會令魚肉霉。同時蒸鹹魚勿下葱絲，否則會臭。

# 魚腸焗蛋

BAKED OMELETTE WITH FISH INTESTINE

## 材 料

| | |
|---|---|
| 鯇魚腸 | 6 副 |
| 白醋 | |
| 冰水 | |
| 中國檸檬 | 2 個 |
| 油炸鬼 | 1 條 |
| 雞蛋 | 4 個（打勻） |
| 陳皮 | 半個 |
| 葱粒 | |
| 芫茜碎 | 適量 |
| 胡椒粉 | |

## 做 法

1. 鯇魚腸放入水盆內，找出腸頭，剪開魚腸。全部魚腸剪開後，將魚腸內壁的污垢刮去，放入粗鹽，揉洗去污垢，沖水。圖 1~2

2. 煲滾水，放入魚腸，不要冚蓋，待黐附魚腸的肥膏差不多溶掉，撈起魚腸放入白醋內（擠些檸檬汁）浸一會去油，撈起。圖 3~5

3. 將魚腸放入冰水（加入檸檬汁），這可以去除醋味和魚腸腥味，並令魚腸爽些，因之前飛水令魚腸腍了。圖 6~8

4. 油炸鬼切片，炸脆。陳皮浸軟，刮去瓤，切幼絲或剁蓉。

5. 瓦缽內，放入魚腸，倒入蛋液，鋪上陳皮絲，放一半油炸鬼；放入已預熱 200℃ 的焗爐內焗 25 分鐘。圖 9~11

6. 取出，放上其餘的油炸鬼、葱粒和芫茜碎，灑上胡椒粉，擠上檸檬汁即可趁熱享用。

## 鼎爺話你知

◆ 用鮫魚腸是這道菜的食材首選，除因鮫魚的魚腸特別大，也因牠是食草的，故腸臟較乾淨。

◆ 將鮫魚腸放入水盆內，鮫魚腸會浮在水盆中，剪魚腸時不會打結，較容易剪開；兼且可以一邊剪開魚腸一邊清洗。

◆ 用滾水淥去魚腸的肥膏時不要冚蓋，否則魚腸會韌。

◆ 浸軟後的陳皮要刮去瓤，否則帶苦味。

◆ 可保留魚肝，味道甘香。

芝麻蝦餅

SESAME SHRIMP CAKE

| | |
|---|---|
| 冰鮮蝦 | 10 兩 |
| 白芝麻 | 適量 |
| 糖 | 少許 |
| 淮鹽 | 少許 |

## 做 法

1. 蝦剝殼後，用鹽醃約 3 分鐘，洗淨。將蝦鋪平放在乾毛巾上，捲起，整卷毛巾放入雪櫃待 2 小時。

2. 從雪櫃取出蝦，將蝦拍扁，用刀背剁成膠狀，再用刀鋒剁幾下。將蝦膠放入碗內，大力撻至起膠，攪勻，下少許糖，攪勻後下少許鹽。圖 1~2

3. 雙手掌心塗少許熟油，取適量蝦膠，將蝦膠搓圓、按薄（炸後的蝦膠會脆些），蘸上芝麻，用大火炸蝦餅至硬身，反轉炸另一面至金黃即可撈起，切成長條，蘸淮鹽享用。圖 3~6

### 鼎 爺 話 你 知

◆ 如用游水鮮蝦製蝦膠，只會感到彈牙爽口，蝦味不濃不甜；想蝦味濃郁和有甜味，就要用冰鮮蝦了，因蝦冰鮮後肉質已糖化。

◆ 宜挑選蝦頭沒有變黑、眼睛沒有下塌的冰鮮蝦。

◆ 用毛巾將蝦捲起再放入雪櫃，可索乾蝦的水分，讓蝦容易起膠兼爽口。

◆ 炸油首選用花生油，因味道較香。

# 煎封羅非魚

SEARED TILAPIA

## 材 料

| | |
|---|---|
| 羅非魚 | 1 條 |
| 薑片 | 2 片 |
| 薑蓉 | 適量 |
| 葱粒 | 適量 |
| 芫茜碎 | 適量 |
| 糖 | 適量 |
| 滾油 | 適量 |
| 豉油 | 適量 |

## 做 法

1. 魚洗淨後，在兩面魚身不對稱的各剋三刀。

2. 下粗鹽醃約 10-20 分鐘，煎時抹去粗鹽。

3. 用大火燒熱鑊下油，用薑片磨一磨鑊底（不易黏鑊兼去腥），轉用中火，放入魚，煎至魚呈微黃後，轉用慢火細煎，可打側略提起平底鑊，煎較難熟的魚頭、較厚肉的魚背。待魚兩面煎至金黃後，用牙籤試熟。圖 1~5

4. 想魚皮脆卜卜，此時倒去鑊內多餘的油，將魚兩面煎至脆皮，上碟。

5. 在魚面放薑蓉、芫茜碎、葱粒，灑上糖，淋滾油，在碟邊下豉油。

6. 要趁熱食，皮脆肉嫩，正！

### 鼎爺話你知

◆ 在魚身兩面不對稱的剠三刀，可以令熱力滲入魚身，容易煎熟。

◆ 先用中火煎魚的用意是，先鎖着肉汁，才改調慢火將魚煎熟，這樣魚肉才會外脆內嫩。

◆ 因為魚頭魚背肉厚，略提起鑊，將油側在一邊，用較多油煎會快些熟，魚肉不會因受火過久而變粗「嗲」。

◆ 不要在魚身淋豉油，否則魚會過鹹；因為在煎魚前已用粗鹽醃過，已有鹹味了。

# 荷芹炒臘味

STIR-FRIED CANTONESE PRESERVED MEAT
WITH SNOW PEAS AND CHINESE CELERY

## 材 料

| | |
|---|---|
| 荷蘭豆 | 半斤 |
| 芹菜 | 1 棵 |
| 菜脯 | 1 兩 |
| 臘肉 | 1/4 條 |
| 臘腸 | 1 條 |
| 薑片 | 1 片 |
| 糖 | 少許 |
| 鹽 | 少許 |

## 做 法

1. 臘腸、臘肉蒸熟，切片。菜脯切幼條。

2. 荷蘭豆洗乾淨，炒前 5 分鐘才去硬邊。

3. 芹菜莖去絲，折斷後再去絲；從芹菜莖中間切開，再切段。

4. 熱鑊下油，用薑片起鑊，至薑片變色（有少許焦），下臘味炒勻，倒入荷蘭豆和菜脯兜勻，下少許糖炒勻，灑入少許鹽兜勻，熄火。最後下芹菜炒勻，上碟。

### 鼎 爺 話 你 知

◆ 荷蘭豆在煮前 5 分鐘才去硬邊，可保留荷蘭豆的香甜。

◆ 臘腸、臘肉在蒸後會較容易切成片狀；如果臘腸只沖洗而不蒸就切片，腸衣會在切時很容易剝落。

# 客家釀豆腐
## STUFFED TOFU IN HAKKA STYLE

| <u>**材**</u> <u>**料**</u> | | <u>**調**</u> <u>**味**</u> <u>**料**</u> | | <u>**蘸**</u> <u>**料**</u> | |
|---|---|---|---|---|---|
| 鯪魚脊 | 3 條 | 糖 | 少許 | 熟油 | 1 湯匙 |
| 硬豆腐 | 2 磚 | 油 | 少許 | 豉油 | 1 湯匙 |
| 蝦米 | 2 兩 | 生粉 | 少許 | 葱粒 | 適量 |
| 冬菇 | 2 兩 | 鹽 | 少許 | | |
| 檸檬葉 | 1 片 | | | | |
| 半肥瘦豬肉 | 4 兩 | | | | |

## 做 法

1. 蝦米浸軟後剁碎；冬菇浸軟後切幼粒；檸檬葉剁成茸；半肥瘦豬肉剁爛。

2. 將鯪魚脊鋪在砧板上，用鐵匙輕力由尾刮向頭，這樣魚肉就會輕易刮出來；如果調轉由頭刮向尾，魚骨就會刮出來，吃時會有骾骨的危險。圖1

3. 魚肉刮出後，再用刀剁爛，再攪撻至起膠。圖2

4. 魚肉、半肥瘦豬肉拌勻，加入蝦米碎、冬菇粒，順時針攪勻，加入檸檬葉茸、少許糖、生粉拌勻至起膠，加入少許油，再攪至均勻，下少許鹽拌勻。

5. 豆腐斜角切成三角形，在豆腐中央割一下，挖起少許豆腐，撲上少許生粉，釀入餡料，煎前再灑少許生粉在魚肉面，煎至餡料金黃，潷去油，再煎，令豆腐更香脆。圖3~8

6. 伴蘸料享用。

### 鼎爺話你知

◆ 可用陳皮茸代替檸檬葉。

◆ 豆腐容易變壞，最好是當天買當天吃，如暫不食用，應用水浸着，烹調時才取出；也可以放入冰格內，做成冰豆腐。

◆ 釀豆腐的餡料煎至金黃時，需要潷去鑊內的油，讓熱力將豆腐烘至香脆。

◆ 釀豆腐煎後加入上湯煮，也非常美味。

# 豆角炒腰膶

STIR-FRIED PORK KIDNEY AND LIVER WITH STRING BEANS

## 鼎爺話你知

◆ 要往相熟、老實的肉檔買豬肝，豬肝以新鮮的為佳，如曾放入雪櫃冷藏，炒後的豬肝一定會硬。

◆ 至於豬腰，不要買脹卜卜的，宜買表面有點乾的，表示沒有啤過水。

◆ 炒豆角時下少許黃砂糖，可以去除豆角的草青味。

## 材料

| 豬腰 | 2 個 |
| 豬肝（膶） | 10 元 |
| 青豆角 | 1 紮 |
| 糖 | 適量 |
| 薑 | 1 片（拍扁） |
| 米酒 | 適量 |
| 鹽 | 少許 |

## 醃料

| 糖 | 1/3 茶匙 |
| 油 | 1/3 茶匙 |
| 生粉 | 1/3 茶匙 |
| 紅葱頭蓉 | 1/3 茶匙 |
| 薑蓉 | 1/3 茶匙 |
| 蒜蓉 | 1/3 茶匙 |
| 鹽 | 1/3 茶匙 |

## 做法

1. 剖開豬腰，切去尿管、脂肪，剝花後切片。將豬腰放入筲箕內，邊揉邊洗，以去除異味，瀝乾水分。圖1~5

2. 豬肝切雙飛，沖水一會，以去除血水和讓豬肝更爽口。撈起，瀝乾水分。圖6~7

3. 豬腰、豬肝用布索乾水分後，一起順序用糖、油、生粉、紅葱頭蓉、薑蓉和蒜蓉醃，拌勻，最後才下鹽。圖8~11

4. 豆角洗淨，切度。圖12

5. 熱鑊下油，倒入豆角兜炒，下糖、薑片炒勻，灒酒，兜炒至5成熟，熄火，撈起豆角。圖13

6. 用原鑊炒豬腰、豬肝，見略熟下豆角兜勻，下糖、鹽，兜勻即可上碟，趁熱享用。圖14~16

梅菜粉葛扣肉

BRAISED PORK BELLY WITH
MEI CAI AND KUDZU

五花腩要刺孔、油泡後，才放入煮腩肉料內烚。

| 材 | 料 | | | 煮 | 腩 | 肉 | 料 | | |
|---|---|---|---|---|---|---|---|---|---|

| 材料 | | 煮腩肉料 | |
|---|---|---|---|
| 新鮮五花腩 | 1 條（約 4 吋寬） | 連皮老薑 | 數片 |
| 粉葛 | 1 個 | 陳皮 | 半個（浸軟刮去瓤） |
| 鹹梅菜 | 2 棵 | 八角 | 2 粒 |
| 甜梅菜 | 1 棵 | 桂皮 | 2 片 |
| 柱侯醬 | 1 湯匙 | 香葉 | 2 片 |
| 老抽（調色用） | | 冰糖 | 適量（須試味） |
| 南乳 | 半磚 | | |

| 上色 | | 汁料 | |
|---|---|---|---|
| 滴珠油 | | 烚五花腩湯 | |

## 做 法

1. 五花腩飛水後，趁熱用滴珠油上色，用叉在豬皮上刺孔。熱鑊下油，以半煎炸形式先炸豬皮那面及略炸肉身。圖 1~4

2. 鍋內放入煮腩肉料，注入清水，煲後試味（如甜味不足，可下多點冰糖），放入五花腩轉用中火煲約 1 小時。

3. 在焗五花腩期間，可切去鹹、甜梅菜菜頭、菜葉，洗淨菜梗，晾乾，切粒。

4. 粉葛去皮，切 1 厘米厚片，蒸 20 分鐘。

5. 用竹籤試五花腩是否已腍，如腍，下柱侯醬和老抽，試味（如口味重的，可下多點柱侯醬），放入南乳和 2/3 鹹、甜梅菜粒，再焗 20 分鐘。

6. 撈起五花腩和梅菜，攤涼，五花腩切厚片，與粉葛相間地排放在蒸碟內，鋪上已煮過的梅菜和餘下未煮過的梅菜，淋上已煮熱的汁料，蒸 20 分鐘即可享用。

## 鼎 爺 話 你 知

◆ 如買不到滴珠油，可將老抽煮熱後，下少許糖，待糖煮溶後再下點油，就可用來上色了。

◆ 五花腩飛水後，用叉在豬皮上刺孔，是防止在泡油時豬皮因過分膨脹而爆炸，令熱油四濺造成危險。

◆ 使用中火燜五花腩，是讓熱力慢慢滲入豬肉內，如果太大火，外面煮至太硬，裏面才剛熟，吃起來就粗糙。

◆ 粉葛如用來煲湯，不要去皮，因有藥用價值。

◆ 優質的梅菜最重要是頭部短，撕出來少筋，必要時可咬一咬嘗味。洗淨後瀝乾水，要像扭毛巾般扭乾，如不夠力，鋪毛巾在枱下吸乾水分才用。

◆ 為甚麼要將梅菜分次烹調？是讓焓五花腩時有梅菜香，但焓過的梅菜香味已褪色，故保留 1/3 未焓的梅菜鋪面，享用時猶有梅菜香。

# 鬼馬牛肉

STIR-FRIED BEEF WITH
DEEP-FRIED DOUGH STICKS

## 材 料

| | |
|---|---|
| 梅頭牛肉 | 半斤 |
| 油炸鬼 | 半條 |
| 芫茜、葱粒 | 適量 |

## 芡 料

| | |
|---|---|
| 糖 | 半茶匙 |
| 蠔油 | 1 湯匙 |
| 生粉水 | 適量 |

## 調 味 料

| | |
|---|---|
| 糖 | 1.5 茶匙 |
| 油 | 1 湯匙 |
| 生粉 | 少許 |
| 薑蓉 | 少許 |
| 蒜蓉 | 少許 |
| 水 | 少許（最後下，可避免牛肉結成一團） |
| 豉油 | 少許 |

▼ 牛肉逆紋切成薄片

▼ 也可以片成薄片

▼ 如牛肉的厚度較薄，可以切成雙飛。

## 做 法

1. 牛肉逆紋切成薄片，順序放入糖、油、生粉、薑蓉和蒜蓉，每次加入都要拌勻，注入少許水，讓牛肉鬆軟，並可避免牛肉結成一團。臨炒牛肉前才下豉油。

2. 油炸鬼切片，翻炸。圖1

3. 熱鑊下油，倒入牛肉，炒至牛肉有點乾時，加入少許水，讓牛肉片再鬆開，炒勻，下一半油炸鬼，倒入芡料，兜勻，上碟。圖2~3

4. 在碟邊放上其餘的油炸鬼，撒上芫茜、葱粒作裝飾。圖4

### 鼎 爺 話 你 知

◆ 醃肉先放糖、油和生粉，後放薑蒜的用意是，糖可令牛肉軟身，油會令牛肉滑一點，而生粉可將牛肉包裹起來，避免薑蒜搶去牛肉的味道。薑蒜不能太多，稍提味即可，但如不下會有腥味。

◆ 醃牛肉千萬不能加鹽，否則會韌如橡筋。

◆ 將脆卜卜的油炸鬼放在碟邊，避免牛肉的水分令油炸鬼變腍。

# 煎牛肉餅

PAN-FRIED BEEF PATTIES

## 材料

| 梅頭牛肉 | 12 兩 |
| --- | --- |
| 檸檬葉 | 4 片 |
| 馬蹄 | 3 粒 |
| 生粉 | 適量 |

## 醃料

| 糖 | 少許 |
| --- | --- |
| 油 | 少許 |
| 豉油 | 少許 |

## 做法

1. 撕去檸檬葉中間的硬梗，切絲或切片。馬蹄去皮，拍扁後略剁。圖1

2. 牛肉切片，用刀背敲鬆，再剁成肉餅，放在大碗內，攪撻一會，下檸檬葉和馬蹄，拌勻，下糖、油，拌勻，下豉油拌勻，醃約 20 分鐘。圖2

3. 用生粉把牛肉搓勻，搓時視乎手感是否「削」，如果感覺較「削」（水分多），可下多點生粉。圖3

4. 將牛肉揑成球狀；熱鑊下油，牛肉球下鑊時按扁，用中慢火煎熟，上碟，以淮鹽或喼汁蘸吃。圖4~8

**鼎 爺 話 你 知**

◆ 切檸檬葉前撕去中間的硬梗，避免享用時影響口感。

◆ 可用陳皮茸代替檸檬葉。

◆ 醃牛肉時不能下鹽，否則牛肉會被「喇死」，肉質會變硬。

◆ 牛肉球不要放入雪櫃內，否則煎時會出水。

# 自家製馬拉盞炒芥蘭

STIR-FRIED KALE IN HOMEMADE BELACAN SAUCE

## 材　料

| | |
|---|---|
| 芥蘭 | 1斤 |
| 自家製馬拉盞 | 適量（視乎喜好，做法參考 p.14） |
| 薑蓉 | |
| 蒜蓉 | |
| 糖 | 各少許 |
| 鹽 | |
| 米酒 | |

## 做　法

1. 芥蘭洗淨，分開菜莖和葉。

2. 熱鑊下油，倒入芥蘭莖，下少許薑蓉和少許蒜蓉，炒至芥蘭有7、8成熟，倒入芥蘭葉，下少許糖炒勻，下鹽，潷酒，兜勻，上碟。

3. 在菜面鋪上馬拉盞，利用芥蘭熱力散發馬拉盞的香氣。

---

### 鼎爺話你知

◆ 因芥蘭莖和葉的煮熟時間不同，故待芥蘭莖有7、8成熟後，才炒芥蘭葉。

◆ 炒芥蘭時下少許糖，可以去除芥蘭的苦澀味。

# 豉磨醬麵

## 鴨燜頭芋

小菜／麵醬磨豉芋頭燜鴨

BRAISED DUCK WITH
TARO IN SOYBEAN SAUCE

## 材　料

| | |
|---|---|
| 米鴨 | 1 隻 |
| 荔蒲芋 | 1 個 |
| 麵豉醬 | 2 湯匙 |
| 磨豉醬 | 2 湯匙 |
| 桂皮 | 5 片（掰成細塊） |
| 香葉 | 5-6 片（掰成細片） |
| 八角 | 5-6 粒（去柄） |
| 薑片 | 5-6 片 |
| 陳皮 | 2/3 個（浸軟後刮去瓤） |
| 冰糖 | 適量（要試味） |
| 老抽 | 少許 |

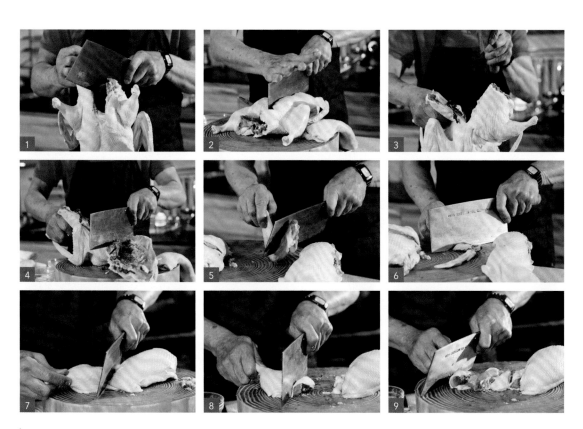

一定要將鴨尾切去，看圖 1。跟着鼎爺的示範，斬鴨沒有難度！

10

11

12

## 做　法

1. 鴨切去尾部，斬大件，凍水下鍋飛水，撈起，瀝乾水分後走油。圖 1~11

2. 芋頭去皮，切成大塊，走油。

3. 燒熱鍋下油，爆香薑片，待薑片半透明後，倒入鴨件、桂皮、香葉、八角、陳皮，慢慢注入平面水，放入適量冰糖。圖 12

4. 待鴨燜至半脸後，倒入芋頭，煮一會待芋頭吸收水分後才下麵豉醬，拌勻後蓋上煲蓋煮至略收水，跟着下一半磨豉醬，試味，若味道不足才下其餘半份磨豉醬，煮一會後倒入少許老抽調色。

5. 用中火再燜 20 分鐘即可享用。

## 鼎 爺 話 你 知

◆ 香料是這菜的靈魂之一，處理香料時用對方法才會事半功倍。八角一定要去柄，否則味道會苦；桂皮和香葉要掰至細塊才出味。還有要凍水下鍋，香料才會出味。

◆ 優質、粉糯的荔蒲芋，切開時芋肉呈紫色筋。

◆ 經走油後的芋頭，燜時不易散開。

◆ 處理鴨時宜切去尾部，因鴨尾很羶，會破壞整個菜的味道。

◆ 因為鴨的脂肪多，所以要先飛水，才再走油。

走油的目的，是令鴨肉緊致，燜時油分會滲入肉內，會更加美味。

◆ 燜鴨放調味的次序好重要，要由淡至濃，所以會先放冰糖，待鴨將熟才下其餘醬料，並要試味，否則如過鹹就不能補救了。

◆ 吃剩了的燜鴨可留在隔天吃，味道更好，但要加添芋頭，因之前的芋頭可能會在加熱時散掉。

# 雞雜炒韭菜花

STIR-FRIED CHICKEN OFFAL WITH
FLOWERING CHINESE CHIVES

## 材料

| | |
|---|---|
| 韭菜花 | 1 紮 |
| 雞雜 | 2 副 |
| 薑蓉 | 1 茶匙 |

## 調味料

糖
油
生粉
鹽

各少許

## 做 法

1. 韭菜花切段，保留花芯。

2. 雞腸撕去肥膏，用生粉醃約 15 分鐘再沖淨，打結，切段，用糖、油和生粉醃一會，臨炒時才下鹽。圖1

3. 雞心片薄片；雞腎切開，刮乾淨雞腎椗，剔花；雞心、雞腎下糖、油和生粉醃一會，臨炒時才下鹽。圖2~3

4. 雞肝切去筋，再片薄，下糖、油、生粉和少許酒醃一會。

5. 熱鑊下油，下薑蓉，倒入雞腎和雞心，炒至雞腎和雞心約 4 成熟時下雞肝，炒至雞肝有 6 成熟時下雞腸，炒至雞腸有 7 成熟時放入韭菜花（先放近根部的），見韭菜花開始軟，下其餘韭菜花和花芯，炒勻，下少許糖及鹽，在鑊邊澆酒，兜一兜即可上碟。圖4

---

### 鼎 爺 話 你 知

◆ 韭菜花的花芯會有少許澀味，炒時下少許糖就可以去除了。

◆ 將雞腎剔花，會易熟、易嚼兼漂亮。

◆ 下少許酒醃雞肝，可辟去腥味。

◆ 炒後的雞腸爽口、彈牙，非常美味。

# 南乳炸雞

DEEP-FRIED CHICKEN MARINATED IN
PRESERVED BEANCURD AND RED TAROCURD

## 材 料

| | |
|---|---|
| 雞 | 1 隻 |
| 生粉 | 1 包 |

## 醃 料

| | |
|---|---|
| 南乳 | 3/4 磚 ⎫ 用少許米 |
| 腐乳 | 1 磚 　 酒調勻 |
| 糖 | 少許 |

## 做 法

1. 雞洗淨，斬塊，倒入醃料醃 1 小時。圖 1~3

2. 雞塊撲上生粉。

3. 猛火燒滾油，下雞塊後收中火，炸好後撈起，約 1 分鐘
   後用大火翻炸，盛起，趁熱享用。圖 4~6

### 鼎 爺 話 你 知

◆ 臨炸前才將雞塊撲上生粉，否則醃雞料會弄濕生粉，
　炸後的雞塊會不夠酥脆。

◆ 撈起翻炸的雞塊時不要熄火，否則油會迫回入炸雞內，
　吃時就會滿口油。

白切雞

SUCCULENT POACHED CHICKEN

## 材料

| | |
|---|---|
| 新鮮雞 | 1 隻（約 2 斤多些，不要放雪櫃） |
| 鹹水草 | 8-10 條 |
| 鹽 | |
| 滾油 | |
| 豉油 | |

## 蘸料

薑蓉

紅葱頭蓉

薑紅葱頭蓉

薑葱蓉

葱蓉

青葱紅葱頭蓉

青葱紅椒圈

黃芥末
（用米酒或白醋調勻，
1 小時後才作蘸料。）

## 做法

1. 光雞洗淨，將雞腳屈入雞腔內，用鹹水草分別綁實雞胸和雞腳，拿着鹹水草將雞放入大火滾水內，收中火滾約 20 分鐘，熄火，浸約 10 分鐘，盛起，浸入冰凍水內至雞凍透。圖 1~2

2. 將雞斬件放在碟上。

3. 每樣蘸料放入小碟內，先放少許鹽，灒滾油，再下豉油（黃芥末不用下豉油）。圖 3~4

## 鼎爺話你知

◆ 光雞約 2 斤餘就最合用，因為太輕就肉薄，而太大則肉會韌。

◆ 建議買二黃雞，即只生過一次蛋，準備生第二次蛋的雞，這類雞身長肉、有雞味、肉夠香；而雞項（未生過蛋的雞），肉質只是嫩滑而沒有雞味。

◆ 用鹹水草綁實雞胸和雞腳，是方便將雞拿進及拿出滾水內。

◆ 將雞浸熟後撈進冰水內，是防止雞肉過熟，以及令皮爽肉滑。

◆ 我今次用酒調稀黃芥末，因為味道會較香。

◆ 用刀背剁薑茸，才會有薑汁，味道夠香。

# 黃酒煮雞

BRAISED CHICKEN IN SHAOXING WINE

## 材 料

雞　　　　　1 隻（切塊）
黑糯米酒　　1 樽（約 500 毫升）
肉薑　　　　約 1.5～2 兩（切片）

## 做 法

1. 熱鑊下油，爆香薑片，待薑片有少許透明時（約 5、6 成熟）放雞塊，不時兜炒，以防止雞塊黐鑊。

2. 兜勻至雞塊有少許轉白色時，沿鑊邊注入黑糯米酒至雞的一半，蓋上鑊蓋，用中慢火燜 20 分鐘即成。

### 鼎 爺 話 你 知

◆ 黑糯米酒有補血、補氣、暖身的功效，客家婦女得悉有孕後，會用黑白糯米自釀糯米酒在產後坐月子時補身。

◆ 不要太早下黑糯米酒，否則會令雞肉泡在酒內過久而不好吃。

◆ 肉薑多用來燜餸煮肉，它的辣味比老薑輕，但肉質較嫩，沒有渣。

糯米釀雞翼

DEEP-FRIED CHICKEN WINGS
STUFFED WITH GLUTINOUS RICE

## 材 料

| | |
|---|---|
| 大隻雞全翼 | 12 隻 |
| 臘味糯米飯 | 約兩碗 |
| 硬身竹籤 | 12 枝 |
| 老抽 | |
| 麥芽糖或燒烤蜂蜜 | 適量 |
| 生粉 | |

## 做 法

1. 雞翼洗淨，切去雞鎚（做其他菜式），中翼起骨。圖 1~5

2. 去骨雞翼釀入糯米飯，用竹籤封口。掃上老抽上色，塗上麥芽糖或燒烤蜂蜜，再撲上生粉。圖 6~12

3. 先用大火燒熱油，轉用中火，放入雞翼慢慢煎至外表金黃內裏熟透即可享用。圖 13~15

### 鼎爺話你知

◆ 釀雞翼時,不要塞得滿滿的,因雞翼
遇熱會收縮,糯米飯會露出來。

◆ 用竹籤封口後,應斬去多餘的竹籤,
避免煎雞翼時竹籤燶掉;但又不可以
斬得太貼近雞肉,避免大啖咬下去時
忘了有竹籤而弄傷口部。

◆ 雞翼掃上老抽上色,煎後的顏色金黃
漂亮,讓人胃口大開。

◆ 雞翼撲上生粉的作用,是令雞翼煎後
有脆邊。

# 三絲蒜蓉
# 蒸三角骨

STEAMED PORK CLAVICLES WITH
DRIED WHITEBAITS,
DRIED SHRIMPS AND SQUID

## 材　料

| | | | |
|---|---|---|---|
| 三角骨 | 1 斤（切件） | 薑蓉 | 1 茶匙 |
| 銀魚乾 | 2 兩 | 紅椒絲 | |
| 蝦乾 | 3 兩 | 蒜片（炸脆） | 適量 |
| 魷魚 | 半條 | 熟油 | |
| 蒜蓉 | 1 湯匙 | | |

## 醃　料

| | |
|---|---|
| 糖 | 1.5 茶匙 |
| 油 | 1 茶匙 |
| 生粉 | 1 茶匙 |
| 豉油 | 1 茶匙 |

## 做　法

1. 銀魚乾、蝦乾分別沖淨後浸軟，蝦乾切絲；魷魚切絲。

2. 將三絲放入大碗內，撈勻，放入排骨，與三絲撈勻。加入糖拌勻，下油拌勻後下生粉，用手指撈勻，感覺糖溶化後加入豉油。圖 1~2

3. 將排骨三絲料放入蒸碟內，撒上蒜蓉、薑蓉和紅椒絲，蒸 20 分鐘，蒸約 5 分鐘時放死氣。圖 3~6

4. 蒸後撒下炸蒜片，淋上熟油，趁熱享用。圖 7~8

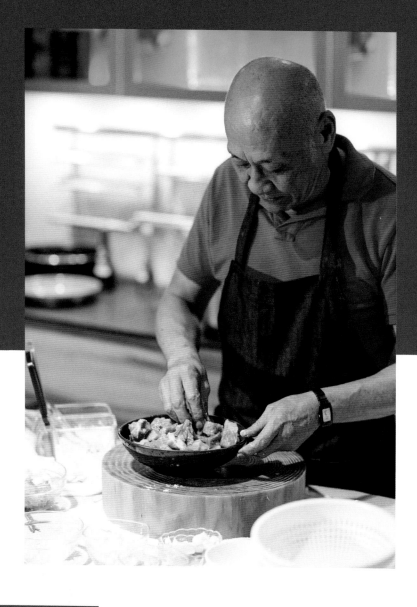

## 鼎 爺 話 你 知

◆ 三角骨又叫三叉骨，我叫它脆脆骨，三角骨用來蒸是香爽甜甘滑，除蒸外，也可以煎炒煮炸燉燜。一隻豬只有少量的三叉骨，要預訂的。

◆ 銀魚乾和蝦乾沖淨浸軟就可以，不要浸過久，否則鮮香味道會消失。

◆ 切鮮魷時，宜切滑的一面，不要用推刀，因如推刀時跣手，會有剕傷的危險，所以要用拖刀。

◆ 用生粉醃肉，可以令肉質鬆軟；而用油可以令肉質軟滑；因為有魚乾和蝦乾的關係，故可放多些糖調味。豉油可帶出食材的鮮味，這是鹽無法比擬的。

◆ 蒸豬肉、雞肉可以下鹽調味，但牛肉就不能了，否則會韌似鐵板。

# 炒五色菜

FIVE-COLOUR VEGGIE STIR-FRY

## 材 料

| | |
|---|---|
| 沙葛 | 2/5 個（切幼條） |
| 木耳絲 | 隨意 |
| 西芹 | 2 枝（撕去硬筋，切幼條）圖1~3 |
| 未炸過的枝竹 | 2 條（浸軟切條） |
| 紅西椒 | 半個（切幼條）圖4 |
| 黃西椒 | 半個（切幼條） |
| 南乳 | 半磚 |
| 腐乳 | 1 磚 |
| 米酒 | 適量 |

## 做 法

1. 南乳、腐乳一起用米酒調勻。

2. 熱鑊下油，下枝竹、沙葛、紅黃椒、西芹炒約 7 成熟，下南乳腐乳料，兜勻，試味，如味道不足，可下少許糖、鹽。圖5

3. 熄火前下木耳絲，兜勻，上碟。圖6

### 鼎 爺 話 你 知

◆ 用沒有椒絲的腐乳。

◆ 五絲不用切得太幼，否則炒後會過腍和「扁塌塌」。

◆ 沙葛可來做菜、醃漬外，也可以生吃，味道清甜、爽脆。

# 生滾鯇魚豬膶粥

GRASS CARP AND PORK LIVER CONGEE

## 材 料

| | |
|---|---|
| 新米 | |
| 水 | |
| 豬膶（肝） | 10元 |
| 鯇魚脊 | 1條 |
| 腐竹 | 1塊 |
| 薑絲 | |
| 葱絲 | |
| 鹽 | |
| 胡椒粉 | 適量 |
| 熟油 | |
| 豉油 | |

## 做 法

1. 煲滾水放入米，轉中火煲至米粒綿滑，放入腐竹煲至溶化。

2. 在煲粥期間，將豬膶和鯇魚脊切雙飛；用粥灼熟豬膶。圖1

3. 取另一瓦煲，放入一隻瓷羹斜靠煲邊，放入薑絲、葱絲，下少許鹽、胡椒粉調味，放入豬膶（預先用粥灼熟）和鯇魚片，淋入滾粥，下少許熟油和豉油，用瓷羹拌勻即可享用。圖2~5

◆ 如將鯇魚脊切成雙飛時有啜刀的情況出現，刀沾少許水就可避免了。

◆ 宜用新米煲粥，因新米水分多，容易煲腍。

◆ 煲粥的米、水比例視乎你想吃稀點還是稠點，一般的米水比例是：1 杯米煲 6 杯水，
但如想稀點就下多些水，想稠的就下少點吧，並要煲久一點。

◆ 煲粥最怕黐底的，其實只要一直保持在翻滾的狀態，別讓粥停在底部便可避免，
但火力也不宜太大，因會滾瀉的。

◆ 放瓷羹在瓦煲內是古老做法，在進食前用瓷羹拌勻，這樣魚片在吃時就不會過熟。

# 生炒臘味糯米飯

STIR-FRIED GLUTINOUS RICE WITH
CANTONESE PRESERVED MEAT

## 材　料

| | |
|---|---|
| 糯米 | 半斤（浸 4 小時，瀝乾水分） |
| 上湯 | 約 1.5 杯 |
| 蝦米 | 1 兩（沖淨，略浸後切粒）圖 1 |
| 冬菇 | 4 朵（浸軟後切粒）圖 2 |
| 瑤柱 | 2 粒（浸軟後拆絲）圖 3 |
| 臘腸 | 1 孖（切粒）圖 4 |
| 膶腸 | 1 孖（切粒） |
| 臘肉 | 1/4 條（切粒）圖 5~6 |
| 豉油 | 少許（取其鮮味） |
| 老抽 | （調色用） |
| 芫茜碎 | 適量 |
| 葱粒 | |

## 做　法

1. 熱鑊下油，用中火炒糯米，至糯米有少許黏鑊時加入上湯（上湯不要沿鑊邊下，要灑在米面），炒勻，冚蓋（焗一會，米才會透心）。

2. 每分半至兩分鐘再開蓋炒勻，下少許油炒，加點上湯，不斷重複，約半小時至飯熟。

3. 先下冬菇、蝦米和瑤柱絲炒勻，約 6 成熟時下臘腸、臘肉。圖 7

4. 炒勻後下豉油，並下老抽調色，倒入芫茜碎、葱粒兜勻即可享用。圖 8~10

## 鼎爺話你知

◆ 糯米浸約 4-5 小時至米粒變白，瀝乾水分才炒；切勿將糯米浸過夜，因米粒太腍，炒時會結成一團。

◆ 臘肉不要切得太大粒，否則臘肉受熱時會收縮，吃時較硬。

◆ 蝦米不用買大隻的，買小一點用刀略剁就成。

◆ 上湯的份量要預備多一點，因每家的爐具火候皆不同。

# 兩面黃煲仔飯

CLAY POT RICE WITH CRACKLING
CRISPY CRUST ON ALL SIDES

## 材　料

| | |
|---|---|
| 絲苗米 | 2 杯（1/3 新米、2/3 舊米） |
| 水 | 2 杯 |
| 臘鴨皮 | 數片 |
| 臘腸 | 1 孖 |
| 膶腸 | 1 條 |

## 甜　豉　油　汁

| | |
|---|---|
| 豉油 | ⎫ |
| 老抽 | ⎬ 煮滾至糖溶化 |
| 糖 | ⎭ |

## 用　具

燒烤網

## 做 法

1. 米洗淨，加入水（米水比例是 1：1），蓋上瓦煲蓋。圖 1~3

2. 用大火煲至將近滾時，轉為中火，見飯面開始收水時，放臘鴨皮在瓦煲邊，臘腸、膶腸放在飯面。圖 4

3. 當聞到有飯焦香，冒出白煙時，將瓦煲稍微傾側，每邊都要烤大約 3 分鐘。圖 5~6

4. 爐火轉到最細。提起瓦煲，火爐上放燒烤網，將瓦煲反轉放（即煲蓋在底，煲底朝面）。當聽到 「答答」聲，並聞到飯焦香時，表示兩面黃已完成。圖 7~16

5. 煮滾甜豉油，淋在飯面，撈勻享用。圖 17~22

21　22

## 鼎爺話你知

◆ 宜買內裏有瓷面的煲仔，做兩面黃
　成功的機會會增大。

23

◆ 煮煲仔飯時一步都不能走開，要金
　睛火眼睇實煲仔飯的變化。

◆ 煮煲仔飯最好是新舊米摻雜，比例
　是 1/3 新米（左，有光澤的）、
　2/3 舊米（右）。如全部用舊米，
　宜加多少許水，因舊米的吸水能力
　高。但一本通書不能睇到老，要看
　看食材是否會出水。多做幾次，就
　會掌握到箇中竅門了。

◆ 放臘鴨皮在瓦煲邊，可讓鴨油滲入
　飯內，更加美味。

◆ 甜豉油和煲仔飯是 perfect match！

◆ 粒粒皆辛苦，以前的人較惜食，
　將普洱茶沖入飯焦內，加芫茜、
　葱，可以消滯兼下火。

# 章魚雞粒有味飯

CHICKEN AND OCTOPUS RICE IN CLAYPOT

## 材 料

| | |
|---|---|
| 絲苗米 | 2 杯（1/3 新米、2/3 舊米） |
| 水 | 2 杯 |
| 雞肉 | （切粒） |
| 冬菇 | （浸軟切粒） |
| 蝦米 | （浸軟切粒） |
| 瑤柱 | （浸軟拆絲） |
| 章魚 | （浸軟切粒） |
| 鮑魚 | （切粒） |
| 急凍中蝦 | （解凍；去腸；切粒） |
| 紅葱頭粒 | |
| 葱粒 | |

雞肉、冬菇、蝦米、瑤柱、章魚、鮑魚、急凍中蝦　適量

## 甜 豉 油

| | |
|---|---|
| 豉油 | |
| 老抽 | 適量 |
| 糖 | |

## 做 法

**1.** 雞肉用少許生粉撈勻，可以令雞肉軟滑，避免在烹調時變得粗韌。

**2.** 砂鍋下米，加水，大火煮至呈蝦眼水時下雞粒、冬菇、蝦米、瑤柱和章魚，拌勻後下鮑魚和中蝦。圖 1~3

**3.** 3 分鐘後收中火，約 10 分鐘後熄火，再焗 5 分鐘即可。

**4.** 在焗飯期間，將紅葱頭粒炸脆，煮熱甜豉油（需試味）。

**5.** 享用時，放上已炸脆的紅葱頭粒、葱粒，倒入豉油汁拌勻，即可趁熱享用。圖 4~6

### 鼎 爺 話 你 知

◆ 處理章魚時，宜撕去薄膜及切去觸鬚上的吸盤，因吸盤很硬、難咀嚼。

◆ 切成粒狀的材料宜大小相若，因熟透的時間較一致。

# 豉油王炒麵

FRIED NOODLES IN SOY SAUCE

## 材　料

| | |
|---|---|
| 炒麵餅 | 2 個 |
| 洋葱 | 1 個（切條） |
| 芽菜 | 4 兩 |
| 葱 | 1 棵（切度） |
| 炒香白芝麻 | 適量 |

## 調　味　料

| | |
|---|---|
| 老抽 | 1 湯匙 |
| 豉油 | 1 湯匙 |
| 糖 | 少許 |

## 做　法

1. 煲滾水，麵餅用熱水弄散，盛起過冷河，瀝乾水後用手弄鬆麵條。

2. 熱鑊下油，倒入洋葱和芽菜各半份略兜炒，下炒麵，用筷子不斷挑鬆；炒麵有少許香味散發出來時，下少許糖調味。圖 1~2

3. 倒入餘下的洋葱和芽菜，加入葱，臨上碟前下老抽調色，豉油調味，盛起，灑上炒香白芝麻。圖 3~4

## 鼎 爺 話 你 知

◆ 將麵條飛水後再過冷河,麵條會較為爽口。過冷河後的麵條一定要瀝乾水分,並弄鬆麵條,就容易炒得成功。圖 5~9

◆ 炒麵前先倒入各半份洋葱和芽菜,才下麵條炒,可減低黐底的機會。另外半份待麵條將熟時才放,那炒麵就有洋葱的香味了。

◆ 炒麵時宜用筷子挑鬆麵條,勿用鑊鏟,因麵條會纏繞鑊鏟,炒時有點「論盡」。

# 綠豆蓮藕陳皮
# 蜜棗乳鴿湯

SQUAB SOUP WITH MUNG BEANS
AND LOTUS ROOT

## 材　料

| | |
|---|---|
| 綠豆 | 4 兩（沖淨） |
| 蓮藕 | 1 節 |
| 陳皮 | 2/3 個（浸軟後刮去瓤） |
| 乳鴿 | 1 隻（洗淨後飛水） |
| 蜜棗 | 2 兩（沖淨） |

## 做　法

**1.** 蓮藕去皮切厚片。

**2.** 煲內放綠豆、陳皮、蓮藕，注入凍水，用大火煲滾，轉用中火，待綠豆開花。

**3.** 加入乳鴿，煲約 40 分鐘，放入蜜棗，再煲約 40 分鐘即成。

### 鼎爺話你知

◆ 試味後，飲用前才在個別的碗內放鹽，因為如整煲湯下鹽調味，湯有剩隔天才飲，湯會有酸味。

# 鱷魚肉川貝燉雞湯

DOUBLE-STEAMED CHICKEN SOUP WITH
CROCODILE MEAT AND CHUAN BEI

## 材　料

| | |
|---|---|
| 川貝 | 半兩 |
| 雪梨 | 1 個（大） |
| 乾鱷魚肉 | 2 兩 |
| 新鮮雞腳 | 10 隻 |
| 雞 | 半隻 |
| 煲湯瘦肉 | 半斤 |
| 火腿 | 適量（愛濃味，可多點） |
| 陳皮 | 半個 |

## 做　法

1. 川貝略舂碎。雪梨切開四份，去芯，留皮。

2. 乾鱷魚肉飛水，以去苦澀味。

3. 雞腳去黃衣，拔去趾甲，洗淨雞腳，將雞腳切半。雞去皮，切大件，飛水。圖 1~3

4. 瘦肉切大件，飛水。火腿飛水，以去澀味。

5. 陳皮浸軟，刮去瓤。

6. 所有材料放入燉盅內，注入過面滾水，用砂紙包裹燉盅燉約 4 小時。

川貝略舂碎便可

## 鼎爺話你知

◆ 這燉湯能舒緩氣管敏感，適宜在寒冷天氣、季節轉換時飲用。

◆ 乾鱷魚肉在藥材舖有售。

◆ 凡煮燉藥材湯，要用刀敲斷雞骨，否則藥會迫入雞骨內，影響功效。

◆ 不用斬去雞腳趾，只需拔去外層的硬趾，否則湯會有膠質，影響口感。

◆ 如買不到大雪梨，可用兩個細雪梨代替。

◆ 用砂紙包裹燉盅，可保存燉湯的香氣；如沒有砂紙，可用微波爐保鮮紙代替。

# 雞蓉粟米魚肚羹

SWEET CORN THICK SOUP WITH
CHICKEN AND FISH TRIPE

## 材 料

| | |
|---|---|
| 雞肉 | 6 兩 |
| 砂爆魚肚 | 2 兩（浸軟） |
| 罐頭粟米粒 | 1 盒 |
| 罐頭粟米蓉 | 1 盒 |
| 瑤柱 | 5 粒（預先用過面水浸泡至軟） |
| 雞蛋白 | 3 個 |
| 鹽 | 少許 |
| 水 | 6 杯 |
| 金華火腿絲（灑面用） | 適量 |

## 芡 料

| | |
|---|---|
| 馬蹄粉 | 2 茶匙 ⎤ 拌勻 |
| 水 | 4 湯匙 ⎦ |

## 做 法

1. 雞肉剁蓉，加入水調勻成稀糊狀。

2. 砂爆魚肚浸軟，切粒。

3. 煲滾水，先下原粒粟米，滾後煮一會，下粟米蓉，待再
   滾下雞蓉，雞蓉要邊下邊攪勻。

4. 不要蓋上煲蓋，待滾後，下魚肚，並加入少許鹽調味；
   待魚肚有少許透明，轉中火，下蛋白拌勻，再勾馬蹄粉
   芡，拌勻，吃時下金華火腿絲。圖 1~4

### 鼎 爺 話 你 知

◆ 因雞肉有膠質，將雞蓉加入水調勻成稀
  糊狀，可避免雞蓉煮時結成一團。

◆ 用馬蹄粉打芡的好處是，就算湯羹攪拌
  過久，也不會出水散開。

◆ 砂爆魚肚的特點是烹調時容易掌握，較
  爽口，不容易起漿。

◆ 讓瑤柱又鬆又軟的竅門？要撕去瑤柱的
  枕。將瑤柱放在盤子內，注入平面水分，
  水剛好蓋過瑤柱，讓瑤柱吸收水分，直
  至所有水分被吸乾；蒸約 30-35 分鐘，
  關火，不要開蓋，再焗 10-20 分鐘才開
  蓋。用多少就捏碎多少用。瑤柱以少裂
  痕為佳，越大的瑤柱越鬆化、越甜，而
  且鮮美。

◆ 打蛋白芡時勿用大火，否則蛋白會變老。

# 芥菜胡椒豬肚湯

PORK TRIPE SOUP WITH PEPPERCORNS
AND MUSTARD GREENS

## 材料

| | |
|---|---|
| 大芥菜 | 1 個 |
| 鹹菜 | 1/3 個 |
| 排骨 | 半斤 |
| 急凍豬肚 | 1 個 |
| 黃豆 | 2 湯匙 |
| 白胡椒粒 | 1/3 兩 |

## 做法

1. 大芥菜逐瓣剝開，洗淨，切大塊。鹹菜用水浸 20 分鐘，撈起。

2. 排骨斬件後飛水。急凍豬肚解凍後，反轉，用粗鹽揉去內裏的肥膏和污垢，清洗乾淨，飛水，切大件。

3. 白胡椒粒用臼春裂。黃豆用清水浸至略軟。

4. 煲內放入排骨、豬肚、鹹菜、大芥菜、黃豆和注入適量水，水滾後加入白胡椒粒，轉中火煲約 1.5 小時。

### 鼎爺話你知

- 要選用菜頭圓、葉要厚，掰開有芥菜香味的大芥菜。

- 白胡椒粒春裂就可，不需要春至全碎。

- 鹹菜用水浸後要咬少許試味，因有些鹹菜會太鹹，所以鹹菜份量不一定要 1/3 個。

- 將排骨飛水，除了去掉血水外，也會讓肉質收縮，味道保留在肉內。圖 1~3

- 加入黃豆同煲湯，湯味有黃豆香兼更有營養。

# 沙參玉竹圓肉螺頭湯

PORK AND CHICKEN SOUP WITH CONCH, SHA SHEN, YU ZHU AND DRIED LONGANS

材料

| | |
|---|---|
| 急凍螺頭（小） | 5 個 |
| 瘦肉 | 6 兩 |
| 雞 | 半隻 |
| 陳皮 | |
| 沙參 | |
| 玉竹 | |
| 淮山 | 請藥材舖 |
| 圓肉 | 執 4 人份量 |
| 杞子 | |
| 蜜棗 | 2 粒 |

## 做法

1. 瘦肉、雞洗淨後分別飛水。圖 1~2

2. 螺頭去厴，剖開，清除腸臟；將螺頭的吸管剝開，去掉污垢；洗乾淨螺頭後飛水。圖 3~8

3. 陳皮浸軟去瓤。

4. 沙參、玉竹、淮山先浸後沖洗；蜜棗、圓肉、杞子略沖淨。

5. 陳皮、瘦肉、螺頭和雞放入鍋，下凍水開大火煲滾，滾後下沙參、玉竹和淮山，用中火煲約 1.5 小時，放入蜜棗和圓肉，改慢火煲 1 小時，最後放入杞子煲半小時即可飲用。

### 鼎爺話你知

◆ 如用螺頭煲湯，我會保留吸管，煲後爽口。

◆ 瘦肉、雞和螺頭一定要先飛水去除血污、肥膩才煲湯，湯才不會濁。

◆ 煲老火湯我會先落硬料，煲至出味後，才下其他易出味的湯料，例如圓肉、杞子等。

# 椰菜瑤柱海皇湯

BRAISED CABBAGE IN GOURMET SEAFOOD SOUP

## 材 料

| 圓椰菜 | 1 個（細） |
|---|---|
| 原粒瑤柱 | 4 粒 |
| 蝦乾 | 1 兩（沖淨略浸） |
| 章魚乾 | 1 隻（洗淨，浸軟後撕去薄膜，切條） |
| 蠔豉 | 8 粒（洗淨略浸） |
| 金華火腿 | 2 粒 |
| 冬菇 | 4 朵（浸軟後切片） |
| 罐頭鮑魚 | 1 隻（切片） |
| 中蝦 | 6 隻（剪去蝦槍、鬚腳） |

## 做 法

1. 略切椰菜底部，剝去外層老衣，用小刀起出中間的芯。圖 1~5

2. 煲滾水，放入瑤柱、蝦乾、章魚乾、蠔豉、金華火腿、冬菇和鮑魚，煲 30-40 分鐘。

3. 放入蝦，不要蓋鍋，待蝦開始熟，才放入椰菜在鍋中央（椰菜的切口向下），如椰菜較細，煲 20 分鐘就行了。圖 6~7

◆ 用小刀起出椰菜中間的芯，因芯太嫩，煲得太久會溶掉，湯會不夠清澈。

◆ 將椰菜的切口向下是讓海味的鮮香滲入菜內，湯和椰菜是精華所在。

◆ 要切去金華火腿外皮黑色的部位，因為會有油臘味。圖8~12

# 龍躉骨大豆芽
# 豆腐番茄湯

GAROUPA BONE SOUP WITH BEAN SPROUTS, TOFU AND TOMATO

## 材 料

| | |
|---|---|
| 龍躉骨 | 1 斤 |
| 大豆芽 | 半斤 |
| 番茄 | 1 個 |
| 板豆腐 | 1 磚（切成方塊） |
| 薑片 | 2 片 |
| 鹽 | 少許 |

## 做 法

1. 煲滾水，放薑片，下龍躉骨，用中大火煲至湯略呈奶白色後，放入大豆芽煲約 30 分鐘。

2. 下番茄（因番茄浮面，用湯杓或筷子將它壓在底），放入豆腐煲約 20 分鐘，下少許鹽調味。

### 鼎 爺 話 你 知

◆ 如用鹹水魚，魚骨不用煎，如用淡水魚魚骨則需略煎，因帶點泥味。

◆ 如想一魚兩味，可同時間炮製西蘭花韭黃炒斑球，將剩下的魚骨用來煲湯。

◆ 在煲邊放上筷子的原因，是蓋上鍋蓋時留一條縫，防止煲湯時滾瀉。

# 腐竹白果
# 雞蛋糖水

BEANCURD SKIN SWEET SOUP
WITH GINGKOES AND EGGS

## 材　料

| | |
|---|---|
| 雞蛋 | 6 個（焓熟） |
| 白果 | 半斤 |
| 腐竹 | 2 片 |
| 雞蛋 | 2 個（拂勻） |
| 冰糖 | 適量 |

## 做　法

1. 煮滾水，下白果，熄火浸 20 分鐘，白果衣自會褪出。圖 1

2. 煲滾水，下冰糖，糖溶後試味，放入白果煲約 10 分鐘；下腐竹，待腐竹溶化後，下焓熟蛋，再下蛋花拌勻，熄火。圖 2~5

# 香蕉木瓜
# 雪耳糖水

WHITE FUNGUS SWEET SOUP
WITH PAPAYA AND BANANA

## 材　料

| | |
|---|---|
| 熟木瓜 | 1 個 |
| 冰糖 | 適量 |
| 香蕉 | 2 條（去皮、切塊）圖 1 |
| 鮮雪耳 | 1 朵（浸軟，去硬蒂，撕成小塊） |
| 杞子 | 約半兩 |

## 做　法

1. 木瓜去皮、去籽、切塊。

2. 將適量水和冰糖煮溶，試味，放入木瓜塊，煮約 3 分鐘，放入香蕉塊，待香蕉邊開始溶，放入杞子和雪耳，再滾後即可享用。圖 2~5

### 鼎爺話你知

◆ 這個懷舊糖水，有滋潤大腸、滑皮膚的功效。

◆ 宜用半生熟的香蕉。

◆ 宜選用瓜身瘦長、籽少的公木瓜煲湯或煲糖水。

◆ 煲這糖水，放木瓜或香蕉沒有一定的先後次序，如木瓜較生，就先放木瓜；如香蕉較生，就先放香蕉。

## TABLE SALT

*\* Refer to p.13 for steps.*

This is the kind of salt that Steve uses on daily basis. But he always makes his own table salt from coarse salt before using.

Fry the coarse salt in a dry wok until white fine powder appears. Finely crush the salt in mortar and pestle. Sieve the salt to remove any big granules.

*pictures 1~6 (upper)*

## SPICED SALT

*\* Refer to p.13 for steps.*

This is used as a dip for deep-fried food such as sesame shrimp cake or deep-fried battered oysters. Steve also makes his own from scratch.

Fry the coarse salt in a dry wok until white fine powder appears. Finely crush the salt in mortar and pestle. Sieve the salt to remove any big granules. Add five-spice powder. Stir well into spiced salt.

*pictures 1~2 (below)*

# HOMEMADE BELACAN SAUCE

* Refer to p.15~16 for steps.

## INGREDIENTS:

80 g Belacan (Malaysian fermented shrimp paste)
6 shallots (finely chopped)
2 cloves garlic (finely chopped)
300 g dried shrimps (soaked in water till soft; finely chopped)
12 dried scallops (soaked in water till soft; broken into shreds)
225 g dried whitebaits (do not rinse)
3 tbsp sugar
3 red chillies (chopped)

## METHOD:

**1.** Deep-fry the dried whitebaits in hot oil until lightly browned. Drain. Let cool and finely chop them. *pictures 1~3, see p.16*

**2.** Prepare the dried shrimps and dried scallops. *pictures 4-5, see p.16*

**3.** Cut Belacan into chunks and put them on a barbecue wire mesh. Grill the Belacan over naked flame to remove the stale and fishy smell. Grill until the Belacan looks dry and fluffy on the outside. Transfer into a bowl and let cool. Add oil and stir into a paste. *see p.15 for steps*

**4.** Heat a wok and add oil. Add half of the garlic and shallot. Put in dried shrimps and dried scallops. Stir until the dried shrimps turn golden. Add whitebaits and brown sugar. Stir until almost dry. Add a little Belacan paste. Toss well. Put in the rest of the Belacan paste, and the remaining garlic and shallot. Toss till almost dry. Add red chillies. Turn off the heat. Spread the sauce evenly on the wok. Cover the lid and let it sit in there for a while. The flavours will be intensified that way. *pictures 6~10, see p.16*

### GRANDPA'S TIPS

◆ *Grilling the Belacan over naked flame helps remove the stale fishy smell. It also fluff up the Belacan so that it combines better with oil and won't stick together in lumps when fried. Make sure you grill the Belacan over the lowest heat possible. It tends to burn easily.*

◆ *If you can't have a wire mesh, you can fry Belacan in a non-stick pan over very low heat.*

◆ *Belacan blocks are available from grocery store specializing in South East Asian food. Do not use Cantonese fermented shrimp paste to substitute.*

## STEAMED CRAB WITH PICKLED PLUMS

*\* Refer to p.20~21 for steps.*

### GRANDPA'S TIPS

- *The digestive tract of the crab is very Cold in nature from Chinese medical point of view and should be removed.*
- *The steaming time depends on the size of the crab.*
- *Crabs are the fattest and most delicious in Lunar January, May and September. This time I use coral crabs for this recipe. They are fleshy and flavourful. They taste best when steamed.*

### INGREDIENTS:

2 coral crabs (about 5 inches wide each)
1 Chinese lemon (squeezed right before serving)

### PLUM SAUCE:

2 pickled Chinese plums
1 tsp sugar
1 tsp grated garlic
1/2 tsp grated ginger
rice wine
caltrop starch
1 tsp oil
salt

### METHOD:

1. To make the plum sauce, de-seed the pickled plums and finely chop them. Add sugar, garlic, ginger. Mix well. Add a dash of rice wine and stir again. Add caltrop starch and oil at last. Mix well. (Caltrop starch and oil will seal in the flavours and stop them from mingling. They should be added after all seasoning is mixed well.) Sprinkle with salt at last and taste it. Add more sugar if too sour.

2. Dress the crab. Pierce two pointy chopsticks from both side between the last two legs all the way to the centre of the crab. The crab's heart will be damaged instantly and it won't struggle and lose its legs when steamed whole. But in this recipe, we will keep on dressing the crab by lifting the carapace. Push the mouthpiece inward and remove the whole digestive tract with it. Remove the sand sac hidden in the roe. Then trim off the two flaps on the shell that cover the roe. These flaps make the heat harder to reach the roe, so that it takes longer to cook through than the rest of the crab. Remove the gills. Cut off the claws. Remove the abdomen by lifting the rear end of the flap. As this is not a "Yellow oil crab" (freshwater crab with yellow fat melted and seeping through its body), you may cut off the ends of its legs. Cut the crab in halves and then quarters. Put a claw along the rim of a chopping board with the plump rounded side up. Crack the claw gently with the flat side of a knife. *pictures 1~10*

3. Arrange the crab pieces on a steaming plate. Make sure the carapaces are placed with the roe side up. Spread the plum sauce over the crab. Steam for 10 minutes. (You don't need to let the steam off halfway as the crab flesh is sliced open and won't retain any stale smell.) *pictures 11~12*

4. Squeeze some Chinese lemon juice over the crabs. Serve.

# STEAMED CROUCHING GAROUPA

*Refer to p.24 for steps.*

## INGREDIENTS:

1 garoupa (about 900 g)
1 pair bamboo chopsticks
shredded ginger
shredded spring onion
sizzling hot oil

### SWEET SOY SAUCE:
**(cooked in a pot till sugar dissolves)**

light soy sauce
sugar

## METHOD:

**1.** Dress the fish and cut its belly from the head to its tail. Rinse well. *pictures 1~2*

**2.** On a steaming plate, put one chopstick on each end. Prop the fish on the chopsticks by lifting both sides of its belly so that it lies on its belly over the chopsticks. Cover the tail with a small plate. *pictures 3~4*

**3.** Steam the fish over high heat for 9 minutes. Open the lid once to let off the steam after it has been steaming for 3 or 4 minutes. Remove fish from steamer and remove the small plate. Carefully remove the chopsticks without damaging the fish. Put shredded ginger on the plate next to the fish. Put spring onion over the fish. Slowly pour sizzling hot oil on the fish from head to tail. (The head takes longer to cook and you may pour more oil on it.) Drizzle some sweet soy sauce on the side of the plate . Serve. *pictures 5~8*

## GRANDPA'S TIPS

◆ *The fish tail takes shorter to cook because it's less fleshy. I cover it with a small plate so that it doesn't get overcooked by steam directly.*

◆ *I butterfly the fish on its belly and put it on chopsticks. That would ensure the fish get heated evenly.*

◆ *Do not open the lid right after the fish is steamed. The fish shrinks when it is cooked. If you open the lid suddenly and the flesh will be cooled drastically and it would shrink even more. It won't taste as good even after you drizzle with some hot oil.*

◆ *Even the best soy sauce has a hint of bitterness. But if you boil it, the bitterness will disappear. For the amount of sugar, it depends on your personal taste.*

# DEEP-FRIED BATTERED OYSTERS

*\* Refer to p.26 for steps.*

### GRANDPA'S TIPS

- *Put pork caul fat in water once you get home. Otherwise, it would melt and disintegrate on its own. Good pork caul fat should be in one whole piece without too many holes on it. You can hold it up well without being too limp or flabby. **picture 1***

- *The essence of this recipe lies in crispiness. Thus, it's essential to blanch the oysters to drive most of the moisture out from the oysters first. Otherwise, the moisture oozing out of the oysters would dampen the battered crust and make it soggy.*

- *Always fry oysters in peanut oil for the crispy crust.*

- *Do not deep-fry too many oyster at one time. Otherwise, the oil temperature will drop suddenly. The oysters will pick up too much oil and taste greasy that way.*

## INGREDIENTS:

2 tubs shelled oysters from the U.S.
1 to 2 sheets pork caul fat
caltrop starch (or flour)
Worcestershire sauce
spiced salt (see p.138 for method)

## FOR RUBBING OYSTERS:

caltrop starch
day-old rice

## SEALING PASTE:

caltrop starch slurry

## DEEP-FRYING BATTER:
## (MIXED TILL LUMP-FREE)

3 tbsp flour
1 tbsp caltrop starch
2 egg whites
water

## METHOD:

1. Rub caltrop starch and day-old rice on the oysters. Rinse well. Blanch in boiling water briefly to partly cook them. Drain and rinse in cold water immediately. Wipe dry with towel. *picture 2*

2. Lay flat pork caul fat on a chopping board. Cut into rectangles (the length should be enough to wrap around the whole oyster).

3. Coat an oyster in caltrop starch. Wrap it in pork caul fat. Seal the seam with caltrop starch slurry. Coat with flour and then dunk it into the deep-frying batter. Deep-fry in oil about 190°C over medium heat. (Do not deep-fry over high heat. Otherwise, the oysters may be undercooked while the crust browns too much.) Deep-fry until lightly-browned. Drain. *pictures 3~9*

4. Turn the heat up and fry the oysters once more. This helps drive any excess oil out of the oysters and crispy up the battered crust. Deep-fry until golden. Drain. Serve the spiced salt and Worcestershire sauce as dips.

# STEAMED COILED WHITE EEL WITH GARLIC, BLACK BEANS AND OLIVES

*Refer to p.31 for steps.*

## INGREDIENTS:

1 white eel (about 2 feet long)
2 dried lotus leaves (soaked in water till soft)
1 tbsp fermented black beans (diced)
1/2 tbsp Chinese pickled olives (diced)
1 tsp finely chopped ginger
3 tsp grated garlic
1 tsp sugar
2 tsp caltrop starch
1 tbsp oil
1/2 tsp salt
1/4 dried tangerine peel
diced spring onion and coriander

## SEASONING:

cooked oil
light soy sauce

## METHOD:

1. Ask the fishmonger to dress the eel for you, without scraping off the slime. Rub coarse salt over the eel. Then scrub with a scouring pad to remove the slime on the eel. Rinse again.

2. Make cuts on the back of the eel repeatedly across the length from head to tail without cutting through the belly *pictures1~3*. Rub caltrop starch and oil over the skin of the eel (not on the flesh). Leave it for 10 minutes.

3. Soak dried tangerine peel in water till soft. Scrape off the pith. Finely shred half of it. Then chop the rest finely. Set aside. In a bowl, mix together fermented black beans, pickled olives, dried tangerine peel, ginger, garlic, caltrop starch, oil and sugar. Sprinkle with a pinch of salt. This is the black bean sauce.

4. Cut off the stem of the lotus leaves (so that the eel won't be propped up when put over lotus leaves). Cover the hole with a piece of lotus leaf. Prick holes on the lotus leaves. Line a bamboo steamer with lotus leaves. Coil the eel and put it over the lotus leaves. Smear the black bean sauce from step 3 over the eel. Cover with another lotus leaf. Cover the lid. *pictures 4~8*

5. Steam over high heat for 12 minutes. Leave the eel in the steamer over the hot water with the lid covered for 3 minutes. That would allow the eel to pick up the sauce further.

6. Drizzle with cooked oil and light soy sauce. Sprinkle with spring onion and coriander. Serve hot. *pictures 9-12*

## GRANDPA'S TIPS

◆ *I prefer dicing the fermented black beans instead of pounding it. That would let their flavour release slowly into the eel without being too overwhelming and in your face.*

◆ *I did not use much ginger in the black bean sauce. Ginger is used to cover up the fishy taste only. If you use too much, the ginger flavour will cover up the garlic flavour and that's not the intent.*

◆ *Half of the dried tangerine peel is shredded while the other half is finely chopped. Chopping is useful in releasing the flavour and aroma to the eel while the rest is shredded for better mouthfeel when you chew it in your mouth.*

◆ *I line the steamer with lotus leaves underneath and over the eel because lotus leaves impart an elegant fragrance. I also did not open the lid once during the steaming process as I pricked holes on the lotus leaves for that purpose.*

◆ *Eel tends to go stale easily. Thus, always buy live eels and cook them as soon as you can after they are slaughtered. Eating eel that has been dressed for a long time may cause food poisoning.*

# BRAISED GAROUPA WITH SHREDDED PORK, MUSHROOMS AND WOOD EAR FUNGUS

*Refer to p.33 for steps.*

## INGREDIENTS:

1 garoupa tail
75 g pork shoulder butt
10 g wood ear fungus
2 dried shiitake mushrooms
1/4 head Zha Cai
(spicy pickle mustard stem)
1 head Chinese celery
1 red chilli
2 slices ginger
coriander

## SAUCE: (MIXED WELL)

1 cup stock
1/2 tbsp dark soy sauce
1 tsp sugar
salt
1 tsp caltrop starch

## METHOD:

1. Make a few cuts on the thickest parts of each side of the fish tail. Coat the fish tail in caltrop starch. Deep-fry in warm oil until half-cooked. *pictures 1~3*

2. Finely shred the pork. Soak wood ear fungus in water till soft. Shred it. Soak shiitake mushrooms in water till soft. Shred them. Rinse Zha Cai and shred it. Tear tough veins off Chinese celery. Finely shred it. De-seed and shred red chilli. Shred the sliced ginger.

3. Add caltrop starch, sugar and oil to shredded pork. Add salt and mix well. Leave it for 10 minutes.

4. Heat a wok and add oil. Stir-fry shredded pork till half cooked. Put in shiitake mushrooms, Zha Cai and wood ear fungus. Toss well. Set aside.

5. Heat the same wok and add oil. Put in the shredded ginger and fish tail. Pour in the sauce. Bring to the boil. Put in the shredded pork mixture from step 4. Cover the lid and simmer for 15 minutes. Add Chinese celery, red chilli and coriander. Cook briefly. Serve. *pictures 4~5*

## GRANDPA'S TIPS

◆ *I make a few cuts on the fish tail so that the heat can reach deeper in the flesh. The fish will cook more quickly.*

◆ *You may check the doneness of the fish tail by inserting a bamboo skewer into the thickest part. If the juices run clear, it is done.*

## INGREDIENTS:

10 frozen medium prawns (about 5 inches long each)
1 green chilli (cut into pieces)
1 red chilli (cut into pieces)
1/2 onion (cut into wedges)
1/2 clove garlic (grated)
1 shallot (grated)
1 tbsp tomato paste
1 tomato (cut into wedges)
sugar
salt
Worcestershire sauce
1 Chinese lemon (to be squeezed over the prawns)

## METHOD:

1. Cut off the feet, rostrums, tails and the heads up to the eyes of each prawn. Make a cut on the back of each prawn between the first and second segment. Make another cut on the back of each prawn between the last and second last segment. Devein by inserting a toothpick in between the third and fourth segment on the back. Do not do it too forcefully. Otherwise, you may break the vein. Then butterfly the abdomen from the first to third segments by cutting with scissors. Do not cut till the tail.

2. Wipe dry the prawns with dry cloth.

3. Deep-fry (or shallow-fry) the prawns in oil until half-cooked. Drain.

4. Heat a wok and add oil. Stir-fry chillies and onion over high heat. Add grated garlic and shallot. Toss well. Add tomato paste and tomato. Put in the prawns and sugar. (Feel free to use more sugar as this sauce is on the sour side.) Toss well and sprinkle with a pinch of salt. Toss again. Add Worcestershire sauce. Stir and turn off the heat. Transfer onto a serving plate. Squeeze lemon juice over it while hot. Serve.

# FRIED MEDIUM PRAWNS IN TOMATO SAUCE

## GRANDPA'S TIPS

◆ I cut off the feet, rostrums, tails and the heads of the prawns up to the eyes because these parts constitute some enclosed pouches that might burst and splatter the oil when fried.

◆ I butterflied the belly of each prawns in the first three segments because those segments have the thickest flesh. Butterflying them make it easier for them to cook through and pick up seasoning.

## OYSTERS WITH ROAST PORK BELLY, GINGER AND SPRING ONION IN CLAY POT

### GRANDPA'S TIPS

◆ I crush the sliced ginger before frying it, so that the flavours go in the oil more easily.

◆ Before I rinse the oysters, I rub caltrop starch and day-old rice on them. They get into the folds and crevices of the oysters and take the dirt with them when rinsed. Do not rub salt on oysters. Otherwise, the oysters will shrink a lot after cooked.

◆ Oysters must be parboiled before used. Otherwise, they tend to give much water and make the dish watery and soupy.

◆ You need to control the cooking time for oysters precisely. They would turn rubbery if overcooked.

### INGREDIENTS:

2 tubs shelled oysters from the U.S.
300 g roast pork belly
3 slices ginger (crushed)
1 shallot
1 clove garlic
1 sprig spring onion (cut into short lengths)
coriander
red chillies (shredded)

### SEASONING:

3 tsp sugar
2 tsp light soy sauce
1 tsp dark soy sauce (for colouring, added last)
rose wine

### FOR RUBBING OYSTERS:

caltrop starch
day-old rice

### METHOD:

1. Rub caltrop starch and day-old rice on the oysters. Rinse and blanch briefly in barely boiling water. Wipe dry with towel or paper towel.

2. Heat a clay pot. Fry sliced ginger, shallot and garlic until fragrant. Put in the roast pork belly and toss until the pork is lightly burned on the rims. Put in the oysters from step 1. Toss until oysters are half cooked. Add sugar. Stir well. Add light soy sauce. Mix again. Add dark soy sauce for colouring and mix well. If you find the colour too light, add a little more. Cover the lid and cook for a while.

3. Put in spring onion and coriander. Cover the lid and pour rose wine slowly along the rim of the lid to give the ingredients a winey aroma. Turn off the heat and serve the whole pot. *see p.39*

4. Garnish with red chillies when you open the lid on the table.

# STIR-FRIED GAROUPA FILLET WITH BROCCOLI AND YELLOW CHINESE CHIVES

*Refer to p.41 for steps.*

## INGREDIENTS:

1 garoupa (about 900 g)
1 broccoli
75 g yellow Chinese chives
(cut into short lengths)
1 small sliced ginger
red chillies (shredded)

## MARINADE FOR FISH:

sugar
cooked oil
caltrop starch
salt

*Sprinkle with a pinch or a dash of sugar, cooked oil, caltrop starch and salt. The amount of seasoning depends on how much flesh is filleted from the fish. But when you marinate fish, always add seasoning in this exact order: sugar, oil, caltrop starch and salt.*

## METHOD:

1. Cut broccoli into florets. Soak them in water with a dash of vinegar and a pinch of salt for 30 minutes. *pictures 1~2* Drain. Put in boiling water and bring to the boil again. Add a pinch of sugar to remove the grassy taste. Cook till the broccoli is half-done. Add a pinch of salt as seasoning. Add cooked oil. Drain and set aside.

2. Fillet the garoupa and cut the flesh into chunks. The fish bones and head can be used to make fish stock *pictures 3~4*. Add sugar, cooked oil and caltrop starch to the fish. Mix well. Add salt.

3. Heat a wok and add oil. Put in the fish and toss well. Add sliced ginger and fry the fish till half-cooked. Put in the older leaves of yellow Chinese chives. Toss briefly. Then put in the rest of the yellow Chinese chives. Toss again. Put in broccoli and stir well. Sprinkle with a pinch of sugar. Toss and sprinkle with a pinch of salt. Toss again and turn off the heat. Transfer to a serving plate. Garnish with shredded red chillies. Serve. *pictures 5~9*

### GRANDPA'S TIPS

◆ *Soaking broccoli in water with vinegar and salt for 30 minutes helps kill germs and bugs, and sterilize the broccoli.*

◆ *After blanching the broccoli, rinse them in cold water immediately. That would keep it bright green without turning yellow.*

◆ *When you marinate any fish, always use cooked oil. Otherwise, it may taste fishy.*

◆ *Whenever you cook with yellow Chinese chives, do not add wine to them. Otherwise, the alcohol would destroy their essential oils and the yellow Chinese chives would taste bland.*

# STEAMED GRASS CARP BELLY WITH SALTED FISH AND DRIED SHRIMPS

*\* Refer to p.44 for steps.*

## GRANDPA'S TIPS

*I used grass carp belly for this recipe for its greasiness. The fish oil will be picked up by the dried shrimps underneath, whereas the grass carp picks up the pungent taste from the salted fish on top. I don't want to brag, but this dish is simply heavenly.*

*Sprinkling some sugar over the salted fish helps cover up the bitterness of the salted fish skin.*

*When you steam salted fish, make sure you put shredded ginger over it before steaming. However, NEVER put shredded ginger over it when steaming fresh fish as the ginger juice would make the flesh soggy. It's also not advisable to steam salted fish with spring onion as it would make the salted fish stink.*

## INGREDIENTS:
1 grass carp belly
38 g large dried shrimps (rinsed; drained)
1 segment salted fish
("Mei-xiang" type; centre segment preferred)
shredded ginger
sugar
red chilli (sliced)

## SEASONING:
cooked oil
light soy sauce

## METHOD:
1. De-bone the salted fish. Discard the belly. Crush the flesh with the flat side of a knife. That would help release its pungent flavour. *pictures 1~2*

2. Put dried shrimps into a steaming plate. Put the grass carp over them. Top with the salted fish. Arrange ginger and red chilli all over. Sprinkle with sugar.

3. Boil water in a steamer over high heat. Steam the fish for 13 to 14 minutes. Lift the lid once to let off the steam after it has been steamed for 3 minutes. *picture 3*

4. Turn off the heat and leave the fish in the steamer for 1 to 2 minutes with the lid covered. That would allow time for the grass carp to pick up flavours from salted fish and dried shrimps.

5. Heat oil till smoking hot. Pour hot oil and soy sauce around the fish on the plate. Serve. *picture 4*

# BAKED OMELETTE WITH FISH INTESTINE

* Refer to p.48~49 for steps.

## INGREDIENTS:

6 sets grass carp intestine

white vinegar

ice water

2 Chinese lemons

1 deep-fried dough stick

4 eggs (whisked)

1/2 dried tangerine peel

diced spring onion

coriander (finely chopped)

ground white pepper

## METHOD:

**1.** Put fish intestine into a basin of water. Find the end of the intestine. Cut open along the length with scissors. Repeat with all fish intestine. Rub salt on the fish intestine vigorously to remove any dirt. Rinse well. *pictures 1~2*

**2.** Boil a pot of water. Put in the fish intestine and do not cover the lid. When most of the fat that attaches on the fish intestine comes off, drain. Transfer fish intestine to white vinegar. Squeeze some lemon juice in it. Leave it briefly to remove the grease. Drain. *pictures 3~5*

**3.** Put the fish intestine into ice water and squeeze some lemon juice in it. This step removes the vinegar taste and make the fish intestine crunchy in texture. (Blanching in water makes the fish intestine soft. The ice helps create a crunch.) *pictures 6~8*

**4.** Slice the deep-fried dough stick thinly. Deep-fry in oil till crispy. Soak dried tangerine peel in water till soft. Scrape off the pith. Finely shred it or finely chop it.

**5.** In an earthenware bowl, lay the fish intestine on the bottom. Pour in whisked egg. Arrange finely chopped dried tangerine peel on top. Put half of the sliced dough stick over. Bake in a preheated oven at 200°C for 25 minutes. *pictures 9~11*

**6.** Remove from oven. Put on the remaining sliced dough stick, spring onion and coriander. Sprinkle with ground white pepper. Squeeze lemon juice on top. Serve hot.

## GRANDPA'S TIPS

◆ *Grass carp intestine is preferred for this recipe because of its hefty size. It's also because grass carp is herbivorous so that its intestine is clean than those omnivorous fish.*

◆ *When you put fish intestine in a basin of water, it would float on top. Then when you cut it open, it won't coil around itself. You also get to wash it while cutting it open.*

◆ *Do not cover the lid when you blanch the fish intestine in boiling water. Otherwise, it would taste tough and rubbery.*

◆ *Scrape off the pith of the dried tangerine peel after soaking it in water. Otherwise, it would taste bitter.*

◆ *You may also put in the fish liver in this recipe. It adds a mineral taste to the dish that some prefer.*

# SESAME SHRIMP CAKE

* Refer to p.52 for steps.

## INGREDIENTS:

375 g frozen shrimps
white sesames
sugar
spiced salt

## METHOD:

**1.** Thaw and shell the shrimps. Marinate with salt for 3 minutes. Rinse well. Lay a towel on the counter and arrange the shrimps one by one on it. Roll the towel up and refrigerate for 2 hours.

**2.** Remove shrimps from the fridge. Crush the shrimps on a chopping board with the flat side of a knife. Then chop it with the back of a knife till minced. Put the minced shrimp mixture into a mixing bowl. Lift it and slap it forcefully back into the bowl until sticky. Stir well. Add a pinch of sugar. Stir well. Add a pinch of salt. *pictures 1~2*

**3.** Grease your hands with cooked oil. Grab a handful of minced shrimp. Roll it round and then flatten it (flat shrimp cakes tend to be crispier after fried). Roll the shrimp cake in a plate of sesames. Deep-fry shrimp cakes in hot oil over high heat until firm. Flip and fry the other side until golden. Drain. Cut into long strips. Serve with spiced salt as a dip. *pictures 3~6*

### GRANDPA'S TIPS

◈ For making minced shrimp, I prefer frozen shrimps. Live shrimps give a crunchy texture when minced, but they don't have much flavour. Frozen shrimps, on the other hand, are much more flavourful as some proteins are broken down into sugars.

◈ Pick frozen shrimps with heads that look the same colour as the body (not darker) and the eyes that are plump.

◈ I roll the shrimps in a dry towel before refrigerating them. The towel helps pick up moisture from the shrimps. The shrimps turn sticky more easily when minced and the shrimp cake tends to be springy.

◈ For deep-frying, it's best to use peanut oil because of its high smoking point and strong aroma.

## INGREDIENTS:

1 tilapia
2 slices ginger
finely chopped ginger
diced spring onion
finely chopped coriander
sugar
sizzling hot oil
light soy sauce

## METHOD:

**1.** Rinse the fish and make three deep cuts on each side at an angle.

**2.** Rub coarse salt on the fish evenly. Leave it for 10 to 20 minutes. Wipe off all salt before frying.

**3.** Heat a wok over high heat. Add oil. Rub the dry wok with a slice of ginger to cover up the fishy taste of the fish, and to make it less likely to burn and stick to the work. Turn to medium heat and slide the fish into the wok. Fry until lightly browned. Turn to low heat and keep on frying. You may tilt the wok to move the hot oil around to fry parts that take longer to cook, such as the head and the back of the fish. Fry till both sides golden. Test for doneness by inserting a toothpick in the thickest flesh. *pictures 1~5*

**4.** If you want the fish skin to be crispy, drain excess oil and keep on frying until crispy. Save on a serving plate.

**5.** Arrange finely chopped ginger, coriander and spring onion on top of the fish. Sprinkle with sugar. Drizzle sizzling hot oil over it. Pour soy sauce around the fish.

**6.** Serve hot. Or the crispy skin may turn soggy.

# SEARED TILAPIA

*\* Refer to p.55 for steps.*

## GRANDPA'S TIPS

◆ *Making three cuts on each side of the fish helps the heat to penetrate into the thicker flesh. That would save you much cooking time.*

◆ *Searing the fish over medium heat to seal in the juices first. Then turn to low heat to cook it through. That's how you end up with fried fish that is crispy on the outside and tender on the inside.*

◆ *The back of the fish is very fleshy and you can focus on cooking this part longer by tilting the wok and moving the oil to submerge this part. The rest of the fish won't be overcooked or dried out.*

◆ *Do not drizzle soy sauce over the fish. Otherwise it would be too salty. As the fish was marinated with salt before fried. It should be salty enough.*

## STIR-FRIED CANTONESE PRESERVED MEAT WITH SNOW PEAS AND CHINESE CELERY

### GRANDPA'S TIPS

◆ *Tear off the tough veins of snow peas 5 minutes before stir-frying. You can retain the sweetness and juices of the snow peas that way.*

◆ *Cantonese preserved pork sausage and pork belly are easier to slice if you steam them first. If you just rinse them without steaming them, the sausage casing is likely to come off when sliced.*

### INGREDIENTS:

300 g snow peas
1 head Chinese celery
38 g dried salted radish
1/4 piece Cantonese preserved pork belly
1 Cantonese preserved pork sausage
1 slice ginger
sugar
salt

### METHOD:

**1.** Steam preserved pork sausage and pork belly until cooked through. Slice them. Set aside. Cut dried salted radish into strips.

**2.** Rinse the snow peas. Tear off the tough veins 5 minutes before stir-frying.

**3.** Tear off tough veins from the Chinese celery. Snap off one end with your hands. Tear off more veins. Cut each stem along the length into halves. Then cut into short lengths. *see p.57*

**4.** Heat a wok and add oil. Stir-fry sliced ginger until it is lightly browned. Add preserved pork sausage and pork belly. Toss till heated through. Add snow peas and dried salted radish. Toss again. Sprinkle with sugar. Stir well. Sprinkle with salt. Toss again. Turn off the heat and put in Chinese celery. Toss and serve.

# STUFFED TOFU IN HAKKA STYLE

* Refer to p.59 for steps.

## INGREDIENTS:

3 dace fillets
2 cubes firm tofu
75 g dried shrimps
75 g dried shiitake mushrooms
1 Kaffir lime leaf
150 g half-fatty pork

## SEASONING:

sugar
oil
caltrop starch
salt

## DIP:

1 tbsp cooked oil
1 tbsp light soy sauce
diced spring onion

## METHOD:

1. Soak dried shrimps in water till soft. Finely chop them. Soak dried shiitake mushrooms in water till soft. Dice finely and set aside. Finely chop Kaffir lime leaf. Finely chop the pork.

2. Lay a dace fillet on the chopping board. Scrape the meat off with a metal spoon from the tail to the head end. On the other hand, if you scrape from the head to the tail end, the fish bones will be scraped off and your guests may choke on bones. *picture 1*

3. Finely chop the dace flesh from step 2. Slap chopped dace on the chopping board repeatedly until sticky. *picture 2*

4. Mix minced dace with chopped pork. Add dried shrimps and shiitake mushrooms. Stir in one direction. Add Kaffir lime leaf, a pinch of caltrop starch and sugar. Stir in the same direction until sticky. Add a little oil. Stir till well incorporated. Sprinkle with salt and stir well.

5. Cut the tofu diagonally into triangles. Scoop out some tofu at the centre of each triangle. Dust with caltrop starch and stuff it with the minced dace filling from step 4. Sprinkle some caltrop starch on the side with minced dace right before frying. Shallow-fry in oil until minced dace turns golden. Drain oil off the wok. Keep on frying to make the tofu crispier. *pictures 3-8*

6. Serve fried tofu with the dip.

## GRANDPA'S TIPS

◆ You may use finely chopped dried tangerine peel instead of Kaffir lime leaf.

◆ Tofu goes stale very easily. It's advisable to serve on the same day you buy it. If you do not plan to cook it right away, soak it in water and drain right before cooking. You may also put it in the freezer to make honeycomb tofu.

◆ When the minced dace filling is fried till golden, I drain the oil off the wok. That would make the tofu crispier.

◆ Fried stuffed tofu can be served as is. Or, you may cook it in stock. It tastes equally great.

## STIR-FRIED PORK KIDNEY AND LIVER WITH STRING BEANS

*Refer to p.62~63 for steps.*

*Refer to p.62~63 for steps.*

### GRANDPA'S TIPS

◆ *Always buy pork liver from reputable butchers you have a good relation with. Pork liver must be from pigs freshly slaughtered on the same day to be tasty. Once refrigerated, it would turn rubbery and hard after cooked.*

◆ *When you shop for pork kidneys, do not pick those that look plump and shiny. Select those with dry and slightly wrinkly membrane on the surface. That's a sign that the pork kidneys aren't plumped up by soaking in water.*

◆ *Adding brown sugar the string beans when stir-frying helps remove the grassy taste.*

### INGREDIENTS:
2 pork kidneys
HK$10-worth of pork liver
1 bundle green string beans
brown sugar
1 slice ginger (crushed)
rice wine
salt

### MARINADE:
1/3 tsp sugar
1/3 tsp oil
1/3 tsp caltrop starch
1/3 tsp finely chopped shallot
1/3 tsp grated ginger
1/3 tsp grated garlic
1/3 tsp salt

### METHOD:
1. Slice open the pork kidneys along the line of symmetry. Trim off the urethra and fat. Make light crisscross cut on the inside of the kidneys. Then slice them. Put them into a strainer. Rinse them under running water while rubbing them to remove any unpleasant smell. Drain. *pictures 1~5*

2. Cut the liver into butterflied slices. Soak in water briefly to remove blood and make the liver crunchy. Drain. *pictures 6~7*

3. Wipe dry both pork kidneys and liver. Marinate them by adding the following in this particular order: sugar, oil, caltrop starch, shallot, ginger and garlic. Stir well. Sprinkle with salt. *pictures 8~11*

4. Rinse the string beans. Cut into short lengths. *picture 12*

5. Heat a wok and add oil. Toss string beans briefly. Add sugar and sliced ginger. Toss again and sizzle with rice wine. Stir-fry until half-cooked. Toss and turn off the heat. Set string beans aside. *picture 13*

6. Heat oil in the same wok. Stir-fry pork kidneys and liver until half-cooked. Put in the string beans and toss well. Add sugar and salt. Toss and save on a serving plate. Serve hot. *pictures 14~16*

# BRAISED PORK BELLY WITH MEI CAI AND KUDZU

* Refer to p.65 for steps.

## INGREDIENTS:

1 piece fresh pork belly (about 4 inches wide)
1 kudzu
2 stems salty Mei Cai (pickled mustard greens)
1 stem sweet Mei Cai (pickled mustard greens)
1 tbsp Chu Hau sauce
dark soy sauce (for colour only)
1/2 cube preserved red tarocurd

## COLOURING:

pearly soy sauce

## STEW STOCK BASE:

a few slices old ginger (skin on)
1/2 dried tangerine peel
(soaked in water till soft; pith scraped off)
2 whole pods star anise
2 pieces cassia bark
2 bay leaves
rock sugar (to be taste)

## SAUCE:

stew stock base

## METHOD:

1. Blanch pork belly in boiling water. Drain. Brush pearly soy sauce all over while still hot. Prick holes on the pork skin with a fork. Heat a wok and put in more oil than shallow-frying, but less oil than deep-frying. Fry the skin side first till lightly browned. Flip to fry the meat side briefly. Drain. *pictures 1~4*

2. Put the stew stock base ingredients into a pot. Add water. Bring to the boil and taste it. Season with rock sugar accordingly if needed. Cook the pork belly in this stock for 1 hour.

3. In the meantime, cut off the roots and the leaves of the Mei Cai. Use only the stems. Rinse well and drain. Dice them.

4. Peel kudzu. Cut into 1-cm thick slices. Steam for 20 minutes.

5. Pierce the pork with a bamboo skewer to see if it is tender enough. If the pork is tender, add Chu Hau sauce and dark soy sauce. Taste it. (You may add more Chu Hau sauce if you prefer.) Put in the preserved red tarocurd and 2/3 of both salty and sweet Mei Cai stems. Cook for 20 more minutes.

6. Remove the pork belly and Mei Cai from the pot. Let cool. Slice the pork thickly. Arrange in a steaming plate in alternate manner with sliced kudzu. Top with Mei Cai from step 5 and the uncooked Mei Cai. Heat the sauce up and drizzle over Mei Cai. Steam for 20 minutes. Serve.

## GRANDPA'S TIPS

- If you can't get pearly soy sauce, you may make your own colouring soy. Heat up some dark soy sauce. Add sugar. Cook till sugar dissolves. Add a dash of oil. Use this mixture to colour the pork.

- After blanching the pork, I prick holes on the pork skin. This is to release any trapped air in the skin that may burst and splatter the hot oil.

- I cook the pork belly over medium heat to slowly cook it through. If you cook it over high heat, the outside would be too tough when the centre is cooked.

- When you make soups with kudzu, do not peel the skin because of its medicinal value.

- Quality Mei Cai has short thick stems. When you tear off a leaf, there shouldn't be many tough veins. You may also have a bite and check if the taste is right. After rinsing and draining, you should wring out the water as if it is a towel. If you find it hard to do, put a dry towel on the counter. Put the Mei Cai over and press the water out before using.

- I cook the Mei Cai in batches. Some Mei Cai is cooked with the pork to give the pork some flavours. Yet, it would be less attractive in colour and blander in taste after the first cooking step. Thus, I save some uncooked Mei Cai to go on top of the pork, so that it tastes better.

# STIR-FRIED BEEF WITH DEEP-FRIED DOUGH STICKS

*\* Refer to p.70 for steps.*

## INGREDIENTS:
300 g beef shoulder butt
1/2 deep-fried dough stick
coriander
spring onion (diced)

## SEASONING:
1.5 tsp sugar
1 tbsp oil
caltrop starch
grated ginger
grated garlic
water (added last)
light soy sauce

## GLAZE: (MIXED WELL)
1/2 tsp sugar
1 tbsp oyster sauce
caltrop starch slurry

## METHOD:

**1.** Cut the beef into thin slices across the grain. Add sugar, oil, caltrop starch, grated ginger and grated garlic in this particular order. Stir after each addition. Add a dash of water to fluff up the beef and avoid the beef sticking together in lumps. Add light soy sauce right before stir-frying.

**2.** Slice the dough stick and deep-fry it in hot oil. Drain. *picture 1*

**3.** Heat a wok and add oil. Put in the beef and toss until the beef looks dry. Add a little water to separate the slices. Toss well. Add half of the sliced dough stick. Pour in the glaze mixture and stir immediately. Toss the beef and dough stick to coat well in the glaze. Save on a serving plate. *pictures 2~3*

**4.** Arrange the remaining deep-fried dough stick along the rim of the plate. Sprinkle with coriander and spring onion. Serve.

## GRANDPA'S TIPS

◆ *When you marinate beef, add sugar, oil and caltrop starch first. Then add ginger and garlic at last. Sugar helps soften the beef. Oil greases the beef to keep it moist. Caltrop starch wraps around the beef so that ginger and garlic won't overpower the beef flavour. Do not use too much ginger and garlic as they are there to complement the beef flavour, not to overpower it. Yet, you can't skip them altogether, as the beef may taste gamey without them.*

◆ *Never marinate beef with salt. Or else, it would be rubbery and extremely dry.*

◆ *I keep some sliced deep-fried dough stick to be placed on the side, so that it remains crispy without picking up the moisture from the glaze or the beef.*

## INGREDIENTS:

450 g beef shoulder butt
4 Kaffir lime leaves
3 water chestnuts
caltrop starch

## MARINADE:

sugar
oil
light soy sauce

## METHOD:

**1.** Rip the centre vein off each Kaffir lime leaves. Finely shred them or slice them. Set aside. Peel water chestnuts. Crush with the flat side of a knife. Coarsely chop them. *picture 1*

**2.** Slice the beef. Then tap with the back of a knife to tenderize it. Finely chop it and put it into a mixing bowl. Lift the chopped beef and slap it a few times into the bowl till sticky. Add Kaffir lime leaves and water chestnuts. Stir well. Add sugar and oil. Stir well. Leave it for 20 minutes. *picture 2*

**3.** Add caltrop starch and mix well. Check if the beef is of a right consistency. If it feels too watery, add a bit more caltrop starch. *picture 3*

**4.** Shape chopped beef mixture into balls. Heat a wok and add oil. Put in the beef balls and press it gently with a spatula. Fry over medium-low heat until cooked through. Serve with spiced salt or Worcestershire sauce. *pictures 4~8*

# PAN-FRIED BEEF PATTIES

*\* Refer to p.72 for steps.*

*\* Refer to p.72 for steps.*

### GRANDPA'S TIPS

- The centre vein of Kaffir lime leaf is very hard and may interfere with the texture of the beef patties.
- You may use finely chopped dried tangerine peel instead of Kaffir lime leaves.
- When you marinate beef, do not add salt. Otherwise, the beef will be rubbery and dry.
- Do not refrigerate the beef balls before pan-frying. Otherwise, water would ooze out of the beef when fried.

# STIR-FRIED KALE IN HOMEMADE BELACAN SAUCE

**INGREDIENTS:**
600 g kale
homemade Belacan sauce
(see p.139 for method)
grated ginger
grated garlic
sugar
salt
rice wine

**METHOD:**
1. Rinse the kale. Separate the stems and the leaves.
2. Heat a wok and add oil. Put in kale stems and add garlic and ginger. Stir until the kale stems are almost cooked through. Put in the kale leaves and sprinkle with a pinch of sugar and salt. Sizzle with wine and toss well. Serve.
3. Spread homemade Belacan sauce over the kale. The remaining heat in the kale will heat up the sauce and bring out its fragrance.

**GRANDPA'S TIPS**

◆ *The leaves and stems of kale cook at different speed. Thus, you should fry the stems till they are almost cooked before putting in the leaves.*
◆ *When you stir-fry kale, you may add a little sugar to remove its bitterness.*

## INGREDIENTS:

1 dressed duck
1 taro
2 tbsp fermented soybean paste
2 tbsp ground bean paste
5 pieces cassia bark (broken into pieces)
5 to 6 bay leaves (broken into pieces)
5 to 6 pods star anise (with stems removed)
5 to 6 slices ginger
2/3 dried tangerine peel
(soaked in water till soft; with pith scraped off)
rock sugar (to be taste)
dark soy sauce

## METHOD:

**1.** Cut off the tail of the duck. Chop into chunks. Put duck pieces into a pot of cold water. Bring to the boil and cook briefly. Drain and wipe dry. Deep-fry the duck in hot oil until lightly browned. Drain. *pictures 1~11*

**2.** Peel and cut taro into large pieces. Deep-fry in hot oil till lightly browned. Drain.

**3.** Heat a wok and add oil. Stir-fry ginger until fragrant and slightly translucent. Put in the duck, cassia bark, bay leaves, star anise and dried tangerine peel. Stir and add enough water to cover. Add a cube of rock sugar. Cover the lid. *picture 12*

**4.** Bring to the boil and turn to medium heat. Cook till the duck is semi-tender. Put in the taro. Cook briefly for taro to pick up the juices. Add fermented soybean paste. Stir well and cover the lid. Cook till the sauce reduces. Add half of the ground bean paste. Stir and taste it. Add all or part of the remaining half of the ground bean paste accordingly. Cook briefly and add dark soy sauce for colour.

**5.** Simmer over medium heat for 20 more minutes. Serve.

# BRAISED DUCK WITH TARO IN SOYBEAN SAUCE

* Refer to p.77 for steps.

### GRANDPA'S TIPS

◆ *Spices are one of the key element of this recipe. Thus, it's essential that you prepare the spices in the right way before cooking. Make sure you remove the stems from the star anise. Otherwise, it may taste bitter. Make sure cassia bark and bay leaves are broken into pieces for quicker infusion of aromas. Finally, you should add spices to cold water for their flavours to infuse in the sauce.*

◆ *Quality taro should be starchy in texture. When you slice it, you should see some purple veins on the cut.*

◆ *The duck is blanched first to make it less greasy. Then it is fried in oil to tighten the muscles so that they would pick up the oil and seasoning when braised. The meat will taste more flavourful that way.*

◆ *When you season the braised duck, it's important to follow an increasing order of seasoning power. Starting with the least overpowering seasoning — rock sugar. Wait till the duck is cooked before adding the rest. And make sure you taste it along the way. If the sauce is too salty, there's nothing you can do to fix it.*

## STIR-FRIED CHICKEN OFFAL WITH FLOWERING CHINESE CHIVES

*\* Refer to p.82 for steps.*

### GRANDPA'S TIPS

◆ *The stamens and stigmas of flowering Chinese chives may taste slightly acrid. You can remove the acridness by adding some sugar when stir-frying them.*

◆ *The crisscross cuts on the chicken gizzards not only shorten the cooking time, but also make them easier to chew and look great.*

◆ *Marinate the chicken liver with a dash of wine to remove the gamey taste.*

◆ *The chicken intestines taste crunchy after cooked. Simply delicious.*

**INGREDIENTS:**
1 bundle flowering Chinese chives
2 sets chicken offal
1 tsp grated ginger

**SEASONING:**
sugar
oil
caltrop starch
salt

**METHOD:**

1. Cut flowering Chinese chives into short lengths. Do not cut off the stamens and pistils.

2. Tear fat off chicken intestines. Add caltrop starch and rub well. Leave it for 15 minutes. Rinse. Tie a knot on it. Cut into sections. Add sugar, oil and caltrop starch. Mix well and leave it briefly. Sprinkle with a pinch of salt right before stir-frying. *picture 1*

3. Slice the chicken heart thinly. Cut open the chicken gizzards and scrape off the hard tendon. Make crisscross cuts on the inside. Put chicken heart and gizzards into a bowl. Add sugar, oil and caltrop starch. Mix well and leave it briefly. Sprinkle with a pinch of salt right before stir-frying. *pictures 2~3*

4. Cut tendons off the chicken liver. Slice thinly. Marinate with sugar, oil, caltrop starch and rice wine. Mix well and leave it briefly.

5. Heat a wok. Add oil. Stir-fry grated ginger until fragrant. Stir-fry chicken gizzards and heart until half-cooked. Put in chicken liver. Stir-fry until half-cooked. Put in the chicken intestines and cook until almost done. Add flowering Chinese chives (put in the ends without flowers first). Cook until flowering Chinese chives looks limp. Put in the rest of the flowering Chinese chives. Stir well. Add sugar. Toss briefly. Add salt. Drizzle rice wine along the rim of the wok. Toss and serve. *picture 4*

# DEEP-FRIED CHICKEN MARINATED IN PRESERVED BEANCURD AND RED TAROCURD

*\* Refer to p.84 for steps.*

## INGREDIENTS:

1 chicken
1 pack caltrop starch

## MARINADE:

3/4 cube preserved red tarocurd ⎤ *mixed together with*
1 cube preserved beancurd ⎦ *some rice wine*
sugar

## METHOD:

**1.** Rinse the chicken and chop into pieces. Add marinade and mix well. Leave it for 1 hour. *pictures 1~3*

**2.** Coat chicken pieces in caltrop starch.

**3.** Heat oil over high heat. Put in the chicken pieces and turn to medium heat. Deep-fry chicken until done. Remove chicken from hot oil and leave it on a strainer for 1 minute. Turn the heat up. Put the chicken back in to fry for the second time. Drain. Serve hot. *pictures 4~6*

---

### GRANDPA'S TIPS

◆ *Always coat the meat in caltrop starch right before you deep-fry it. Otherwise, the caltrop starch will pick up the moisture and the meat won't be crispy on the outside.*

◆ *Do not turn the heat off before all chicken pieces are removed from the oil after the second deep-frying step. Otherwise, the chicken will pick up more oil and become very greasy.*

# SUCCULENT POACHED CHICKEN

*\* Refer to p.88 for steps.*

## GRANDPA'S TIPS

◆ *A dressed chicken around 1.2 kg would be perfect for this recipe. A smaller chicken would not be fleshy enough. A bigger chicken may be tougher in texture.*

◆ *I prefer "Double-yellow chicken" for this recipe. That means the chicken has laid eggs once and is about to lay eggs the second time. Such chicken is fleshy and flavourful. On the other hand, pullets that have never laid eggs have tender flesh, but it is less flavourful.*

◆ *I tie straw strings on the chicken breast and legs so that I can dip the chicken into hot water and remove it conveniently.*

◆ *After poaching the chicken, I dunk it into ice water to crisp up the skin and to stop it from being overcooked by the remaining heat in the meat.*

◆ *I add rice wine to yellow mustard to elevate its aroma.*

◆ *I chop the ginger with the back of a knife. That would partly juice the ginger and the dips will taste gingery.*

## INGREDIENTS:

1 freshly slaughtered chicken
(1.2 kg or slightly more; not refrigerated)
8 to 10 straw strings
salt
sizzling hot oil
light soy sauce

## DIPS:

finely chopped ginger
finely chopped shallot
finely chopped ginger and shallot
finely chopped ginger and spring onion
finely chopped spring onion
finely chopped shallot and spring onion
sectioned spring onion and sliced red chillies
yellow mustard (mixed with rice wine or white vinegar; let sit for 1 hour before serving)

## METHOD:

1. Rinse the chicken. Tuck the feet into the cavity. Tie straw strings securely on the chicken breast and legs. Dip the chicken into vigorously boiling water by holding the straw strings. Turn to medium heat and boil for 20 minutes. Turn off the heat and let the chicken stay in the hot soup for 10 more minutes. Remove and dip into ice water until completely cooled. *pictures 1~2*

2. Chop the chicken into pieces and arrange on a serving plate.

3. Put each dip into a small dish. Sprinkle with a pinch of salt. Sizzle with hot oil. Add a dash of light soy sauce to all dips except yellow mustard. *pictures 3~4*

# BRAISED CHICKEN IN SHAOXING WINE

**INGREDIENTS:**

1 chicken (chopped into pieces)
1 bottle black glutinous rice wine (500 ml)
57-75 g chunk meaty ginger (sliced)

**METHOD:**

**1.** Heat a wok and add oil. Stir-fry ginger till fragrant. Wait till the ginger turns slightly translucent. Put in the chicken. Toss from time to time, so that it won't burn and stick to the wok.

**2.** When the chicken is half cooked (turns white), pour in half of the black glutinous rice wine along the rim of the wok. Cover the lid and cook over medium-low heat for 20 minutes. Serve.

### GRANDPA'S TIPS

◆ Black glutinous rice wine is believed to promote blood cell regeneration, strengthen Qi (vital energy) flow and warm the body. Hakka women usually start making rice wine with both black and white glutinous rice when they get pregnant, so that they can consume home-brewed rice wine to recuperate after childbirth.

◆ Do not put in the glutinous rice wine too early. Otherwise, the chicken will be cooked in alcohol for too long and won't taste good.

◆ Meaty ginger is usually used in braised recipe. It is not as spicy as old ginger and also much less fibrous. It has tender flesh.

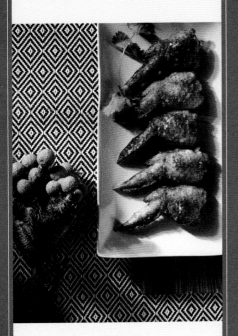

## DEEP-FRIED CHICKEN WINGS STUFFED WITH GLUTINOUS RICE

*Refer to p.93~94 for steps.*

### GRANDPA'S TIPS

◆ *When you stuff the chicken wings with glutinous rice, do not overstuff them. The chicken meat would shrink when cooked. If you overstuff them, the filling may leak.*

◆ *When you seal the seam with a bamboo skewer, trim any excess off the skewer. Otherwise, the exposed skewer may burn when you deep-fry the chicken wings. But you shouldn't trim it too close to the flesh either. Otherwise, your guests may not know there is a skewer and may bite on it and hurt themselves.*

◆ *Brushing dark soy sauce on the chicken wings is purely for presentation. It gives the wings a golden colour which is appetizing.*

◆ *Coating the wings in caltrops starch helps create crunchy bits along the edges of the skin.*

**INGREDIENTS:**
12 large whole chicken wings
2 bowls stir-fried glutinous rice with
Cantonese preserved meat (see recipe on p.168)
12 bamboo skewers
dark soy sauce
maltose (or honey)
caltrop starch

**METHOD:**
1. Rinse the chicken wings. Cut off the drumettes (save it for other recipes). De-bone the mid-joint segment. *pictures 1~5*
2. Stuff the mid-joint segment with cooled glutinous rice. Seal the seam with a bamboo skewer. Brush some dark soy sauce on the skin evenly. Then spread maltose or honey all over. Coat lightly in caltrop starch. *pictures 6~12*
3. Heat oil over high heat in a fryer or wok. Turn to medium heat and deep-fry the stuffed chicken wings until golden on the outside and cooked through. Drain. Serve. *pictures 13~15*

# STEAMED PORK CLAVICLES WITH DRIED WHITEBAITS, DRIED SHRIMPS AND SQUID

*Refer to p.96 for steps.*

## INGREDIENTS:

600 g pork clavicles
(chopped into bite-size pieces)
75 g dried whitebaits
115 g large dried shrimps
1/2 fresh squid
1 tbsp grated garlic
1 tsp finely diced ginger
red chilli (shredded)
sliced garlic (deep-fried till crispy)
cooked oil

## MARINADE:

1.5 tsp sugar
1 tsp oil
1 tsp caltrop starch
1 tsp light soy sauce

## METHOD:

1. Rinse dried whitebaits and dried shrimps separately. Soak them in water till soft separately. Finely shred the dried shrimps. Finely shred the squid.

2. Put dried whitebaits, dried shrimps and squid into a mixing bowl. Mix well. Put in the pork clavicles and mix again. Add sugar and stir well. Add oil and stir again. Add caltrop starch. Stir and rub with your hands. Add light soy sauce when you can't feel the granules of the sugar anymore. *pictures 1~2*

3. Arrange the resulting mixture on a steaming plate. Sprinkle with grated garlic, ginger and red chilli. Steam for 20 minutes. Lift the lid once to let the steam off after it has been steamed for 5 minutes. *pictures 3~6*

4. Sprinkle with deep-fried sliced garlic. Drizzle with cooked oil. Serve hot. *pictures 7~8*

## GRANDPA'S TIPS

◆ Pork clavicle is a fine cut of meat. When steamed, it taste flavourful with a crunchy and velvety texture. Besides steamed, it also works great when fried, stir-fried, deep-fried or braised. Yet, every pig has only two clavicles and this cut comes in strictly limited quantity. You may have to order from your butcher in advance.

◆ Just rinse and soak the dried whitebaits and dried shrimps in water till soft. Do not soak them for too long. Otherwise, they may taste bland and won't add to the flavours of the pork.

◆ When you slice a squid, try to slice on the smooth side of it with a pulling action. If you slice it with a pushing action, the knife may slip and you may cut yourself.

◆ Marinating meat with caltrop starch helps keep the meat moist and tender. Oil also works in similar ways to keep the meat soft. Yet, both dried whitebaits and dried shrimps are salty to begin with, you'd need more sugar than usual in the marinade. I prefer seasoning meat with light soy sauce over salt because its soybean aroma brings out the natural sweetness and flavours of ingredients.

◆ When you steam pork or chicken, you may season it with salt. But never season beef with salt. Otherwise, it'd turn rubbery and tough.

# FIVE-COLOUR VEGGIE STIR-FRY

*\* Refer to p.99 for steps.*

## INGREDIENTS:

2/5 yam bean (cut into thin strips)
wood ear fungus (cut into thin strips)
2 celery stalks (with tough veins torn off; cut into thin strips) *pictures 1~3*
1/2 red bell pepper (cut into thin strips) *picture 4*
1/2 yellow bell pepper (cut into thin strips)
2 dried beancurd sticks (not fried; soaked in water till soft; cut into thin strips)
1/2 cube fermented tarocurd
1 cube fermented beancurd
rice wine

## METHOD:

**1.** Mix fermented tarocurd and fermented beancurd in a small bowl. Thin it out with rice wine. Mix well.

**2.** Heat a wok. Add oil. Stir-fry beancurd sticks, yam bean, bell peppers and celery. Toss until half-cooked. Pour in the tarocurd mixture from step 1. Toss to coat well. Taste it. If it needs more seasoning, add a little salt and sugar. *picture 5*

**3.** Put in the wood ear fungus. Toss again. Serve. *picture 6*

### GRANDPA'S TIPS

◆ *For this recipe, choose plain fermented beancurd without chilli.*

◆ *Do not shred the five vegetables too finely. Otherwise, they would overcook easily and turn soggy.*

◆ *Yam bean is a great ingredient that tastes sweet and crunchy. You may pickle it or stir-fry it. You may even eat it raw.*

## INGREDIENTS:

new rice
water
HK$10-worth of pork liver
1 grass carp fillet
1 sheet beancurd skin
ginger (finely shredded)
spring onion (finely shredded)
salt
pepper
cooked oil
light soy sauce

## METHOD:

**1.** Boil a pot of water. Put in the rice. Bring to the boil over high heat. Then turn to medium heat and cook until the rice grains turn mushy and the water turns thick without the lid. Add beancurd skin. Cook until it breaks down completely.

**2.** While the congee is boiling, cut pork liver and grass carp fillet into butterflied slices. Parboil the pork liver in the boiling congee briefly. Set aside. *picture 1*

**3.** Put a ceramic spoon into a clay pot. Put in shredded ginger and spring onion. Sprinkle with salt and ground white pepper. Put in the parboiled pork liver and raw grass carp. Pour over some boiling congee. Drizzle with cooked oil and light soy sauce. Stir with the ceramic spoon. Serve. *pictures 2~5*

# GRASS CARP AND PORK LIVER CONGEE

*\* Refer to p.101 for steps.*

### GRANDPA'S TIPS

◆ You may find the knife sticking to the fish when you cut the grass carp into butterflied slices. Just dip the knife in water and it won't stick as much.

◆ I prefer making congee with new rice because new rice has higher moisture content and it tends to break down and turn mushy more quickly than old rice.

◆ The ratio between rice and water depends on your preferred congee consistency. As a general guideline, add 6 cups of water to 1 cup of rice. If you like your congee thinner, use more water. If you want your congee thicker, use less water.

◆ Congee is notorious for getting burnt easily. The trick to avoid burning is to keep it vigorously boiling all the time. Do not let the rice grain sink to the bottom throughout the cooking process. Do not turn the heat too high either as it tends to boil over easily.

◆ Putting a ceramic spoon in a clay pot before pouring in congee is a traditional way of serving. Before eating, each guest stirs his own congee with a spoon to cook through the fish with the hot congee. That way, the fish won't be overcooked.

# STIR-FRIED GLUTINOUS RICE WITH CANTONESE PRESERVED MEAT

* Refer to p.104~105 for steps.

## INGREDIENTS:

300 g glutinous rice (soaked in water for 4 hours; drained)
1.5 cups stock
38 g dried shrimps (rinsed; soaked in water till soft; diced) *picture 1*
4 dried shiitake mushrooms (soaked in water till soft; diced) *picture 2*
dried scallops (soaked in water till soft; broken into shreds) *picture 3*
2 Cantonese preserved pork sausages (diced) *picture 4*
2 Cantonese preserved pork liver sausages (diced)
1/4 piece Cantonese preserved pork belly (diced) *pictures 5~6*
light soy sauce
dark soy sauce (for colouring)
coriander (finely chopped)
spring onion (diced)

## METHOD:

**1.** Heat a wok and add oil. Stir-fry glutinous rice over medium heat. When it starts to dry out and turns sticky, add some stock. (Do not add along the rim of the wok. Pour at the centre.) Keep stirring and cover the lid (to let the steam cook through the rice).

**2.** Open the lid and stir the rice every 1.5 or 2 minutes. Add a little oil and some stock. Repeat this step for about 30 minutes until the rice is cooked through.

**3.** Stir-fry shiitake mushrooms, dried shrimps and dried scallops till half-cooked. Add preserved pork sausages and pork belly. Toss well. *picture 7*

**4.** Add light soy sauce for flavour and dark soy sauce for colour. Mix well. Sprinkle with coriander and diced spring onion. Toss again. Serve. *pictures 8-10*

---

### GRANDPA'S TIPS

◆ *Before stir-frying glutinous rice, always soak it in water for 4 hours until the grains turn white. Then drain well before use. Do not soak the rice overnight. Otherwise, it would pick up too much moisture and may turn into one gooey mess when fried.*

◆ *When you dice the Cantonese preserved pork, do not dice it too big. It tends to shrink a bit after fried and it may be too tough to chew.*

◆ *You don't need large dried shrimps for this recipe as it will be chopped anyway. Save them for other uses. For this recipes, use small ones and they take less time to chop.*

◆ *Every stove differs in power and heat. Thus, it's mandatory to prepare more stock than you think you need, so that you won't run out.*

## INGREDIENTS:

2 cups jasmine rice (1/3 new rice plus 2/3 old rice)

2 cups water

a few pieces skin of Cantonese preserved duck

2 Cantonese preserved pork sausage

1 Cantonese preserved pork liver sausage

### SWEET SOY SAUCE:
### (BOILED TILL SUGAR DISSOLVES)

light soy sauce

dark soy sauce

sugar

## UTENSILS:

barbecue wire mesh

## METHOD:

**1.** Rinse the rice and put in a clay pot. Add water (ratio of rice to water is 1:1 by volume). Cover the lid. *pictures 1~3*

**2.** Bring to the boil over high heat. Turn to medium heat. When there is no water on top of the rice, put preserved duck skin on the rim of the pot over the rice. Put pork sausage and pork liver sausage at the centre. *picture 4*

**3.** Cook until there is a mild burning smell and white smoke comes out of the pot. Tilt the clay pot to cook its vertical sides. Make sure you rotate the pot to cook evenly. Cook each spot for about 3 minutes. *pictures 5~6*

**4.** Turn to the lowest heat. Remove the clay pot and put a wire mesh over the stove. Flip the clay pot upside down (so that the lid is on the stove directly). Keep on heating till you hear the pot clicks and smell the rice mildly burnt. Turn off the heat. *pictures 7~16*

**5.** Serve the whole pot. Drizzle the rice with sweet soy sauce. Mix well. Serve. *pictures 17~22*

# CLAY POT RICE WITH CRACKLING CRISPY CRUST ON ALL SIDES

*\* Refer to p.108~109 for steps.*

## GRANDPA'S TIPS

◆ *Get yourself a clay pot with ceramic glaze on the inside. You have much better chance of pulling off this recipe this way.* **picture 23**

◆ *When you make clay pot rice, you must stay close to the stove all the time and observe the way the rice changes.*

◆ *For the best result, mix 1/3 new rice with 2/3 old rice for this recipe. If you use old rice only, you'd need more water because old rice is dryer and picks up more water. But of course, this is a general guideline only. If the rice is topped with ingredients that give much water when cooked, you may not need to add more water. All in all, practice makes perfect. Make the same dish a few times and you'd get to learn the tricks therein.*

◆ *I put the skin of preserved duck on the rim of the rice so that the duck fat would run into the rice and add a lovely flavour.*

◆ *Sweet soy sauce is the perfect match with clay pot rice.*

◆ *People used to cherish food a lot and would not throw anything edible in the garbage. They usually added Pu Er tea to the crispy rice crust and sprinkled coriander and spring onion on top and serve it as another course. Legend has it that it aids digestion and dissipates Fire in the body.*

# CHICKEN AND OCTOPUS RICE IN CLAYPOT

*\* Refer to p.111 for steps.*

## GRANDPA'S TIPS

◆ *After soaking the dried octopus in water to rehydrate it, peel off the skin and cut off all the suckers on its arm. The suckers are too hard to chew.*

◆ *Try to dice all ingredients into pieces of similar size, so that they'd cook through at the same time.*

## INGREDIENTS:
2 cups jasmine rice (1/3 new rice plus 2/3 old rice)
2 cups water
chicken meat (diced)
dried shiitake mushrooms (soaked in water till soft; diced)
dried shrimps (soaked in water till soft; diced)
dried scallops
(soaked in water till soft; broken down into shreds)
dried octopus (soaked in water till soft; diced)
canned abalone (drained; diced)
frozen medium prawns (thawed; deveined; diced)
shallot (diced)
spring onion (diced)

## SWEET SOY SAUCE:
light soy sauce
dark soy sauce
sugar

## METHOD:
**1.** Add caltrop starch to chicken. Mix well. It helps keep the chicken moist and tender. That would retain the juices and prevent the chicken from getting tough.

**2.** Put rice into a clay pot. Add water. Bring to the gentle simmer over high heat. Put in diced chicken, shiitake mushrooms, dried shrimps, dried scallops and dried octopus. Stir well. Put in abalone and prawns. *pictures 1~3*

**3.** Cook over high heat for 3 minutes. Turn to medium heat and cook for 10 more minutes. Turn off the heat and cover the lid. Leave it for 5 minutes.

**4.** While the rice is cooking, deep-fry the diced shallot until crispy. Heat the sweet soy sauce and taste it to adjust the amount of sugar used.

**5.** Before serving, arrange deep-fried shallot and spring onion over the rice. Drizzle with the sweet soy sauce. Stir well and serve hot. *pictures 4~6*

# FRIED NOODLES IN SOY SAUCE

**INGREDIENTS:**
2 bundles noodles for stir-frying
1 onion (cut into strips)
150 g mung bean sprouts
1 sprig spring onion (cut into short lengths)
toasted sesames

**SEASONING:**
1 tbsp dark soy sauce
1 tbsp light soy sauce
sugar

**METHOD:**

1. Boil a pot of water and cook the noodles until they scatter. Drain and rinse in cold water. Drain and fluff up the noodles with your hands.

2. Heat a wok and add oil. Add half of the onion and half of the bean sprouts. Toss briefly. Put in the noodles. Stir and fluff them up in the wok with chopsticks. When you smell the fragrance of the noodles, sprinkle with some sugar. *pictures 1~2*

3. Put in the remaining onion and bean sprouts. Add spring onion. Sprinkle with light soy sauce for flavour and dark soy sauce for colour right before plating. Save on a serving plate and sprinkle with toasted white sesames. Serve. *pictures 3~4*

### GRANDPA'S TIPS

◆ *I blanched the noodles first and rinse them in cold water. The noodles would have a crunchy texture that way. After rinsing the noodles, make sure you drain them very well and fluff them up with your hands. They are a lot easier to handle this way when stir-fried. pictures 5~9*

◆ *I toss half of the onion and bean sprouts in the wok before I put in the noodles. The onion and bean sprouts would separate the noodles from the hot wok so that noodles are less likely to burn and stick to the wok. After the noodles are cooked through, I put in the rest of the onion and bean sprouts so that they won't be overcooked. They also make the noodles more aromatic.*

◆ *When I make fried noodles, I always toss them with chopsticks instead of a spatula. The spatula is entangled with noodles very easily, making them harder to toss and mix well.*

# SQUAB SOUP WITH MUNG BEANS AND LOTUS ROOT

**INGREDIENTS:**

150 g mung beans (rinsed)
1 segment lotus root
2/3 dried tangerine peel
(soaked in water till soft; pith scraped off)
1 squab (rinsed; blanched in boiling water)
75 g candied dates (rinsed)

**METHOD:**

**1.** Peel the lotus root. Cut into thick slices.

**2.** Put mung beans, dried tangerine peel, lotus root and cold water into a pot. Bring to the boil over high heat. Turn to medium heat and boil until the mung beans are mushy and break down.

**3.** Put in the squab. Boil for 40 minutes. Add candied dates. Boil for 40 minutes. Serve.

### GRANDPA'S TIPS

◆ *It's advisable to taste the soup first and season it with salt in individual serving bowls. If you season the whole pot altogether and have leftover, it may turn sour the next day.*

## INGREDIENTS:

19 g Chuan Bei
1 large Ya-li pear
75 g dried crocodile meat
10 fresh chicken feet
1/2 chicken
300 g lean pork
Jinhua ham to taste
(add more if you prefer more flavourful soup)
1/2 dried tangerine peel

## METHOD:

1. Crush Chuan Bei coarsely with mortar and pestle. Set aside. Cut pear into quarters. Core them, but keep the skin on.

2. Blanch dried crocodile meat in boiling water to remove bitterness.

3. Peel the yellow skin off chicken feet. Rip off the nails. Rinse well and chop each into halves. Set aside. Skin the chicken and chop into chunks. Blanch in boiling water. Drain. *pictures 1~3*

4. Cut lean pork into chunks. Blanch in boiling water. Drain. Blanch Jinhua ham in boiling water to remove acridness.

5. Soak dried tangerine peel in water till soft. Scrape off the pith.

6. Put all ingredients into a double-steaming pot. Add enough boiling water to cover all ingredients. Wrap the pot in mulberry paper. Leave the pot in a simmering water bath for 4 hours. Serve.

# DOUBLE-STEAMED CHICKEN SOUP WITH CROCODILE MEAT AND CHUAN BEI

*\* Refer to p.119 for steps.*

## GRANDPA'S TIPS

◆ *This soup alleviates discomfort related to sensitive respiratory tract. Serve it in winter time and when seasons change.*

◆ *When you make chicken soup with herbal ingredients, make sure you crack all bones of the chicken. Otherwise, the potent chemicals may be drawn into the bones and that undermines the medicinal value of the soup.*

◆ *Do not chop off the toes of the chicken feet. You just need to rip off the nails. Otherwise, the soup will turn gelatinous and gooey.*

◆ *You can get dried crocodile meat from Chinese herbal stores.*

◆ *If you can't get large Ya-li pears, you can use two small ones instead.*

◆ *I seal the double-steaming pot in mulberry paper to retain the aroma of the soup. If you can't get mulberry paper, use microwave-safe cling film instead.*

# SWEET CORN THICK SOUP WITH CHICKEN AND FISH TRIPE

*\* Refer to p.122 for steps.*

## INGREDIENTS:

225 g chicken meat
75 g sand-puffed fish tripe (soaked in water till soft)
1 pack sweet corn kernels
1 pack cream style sweet corn
5 dried scallops (soaked in water till soft)
3 egg whites
salt
6 cups water
shredded Jinhua ham (as garnish)

## THICKENING GLAZE: (MIXED WELL)

2 tsp water chestnut starch
4 tbsp water

## METHOD:

**1.** Finely chop the chicken. Add water and stir into a thick paste.

**2.** Soak the fish tripe in water till soft. Dice it.

**3.** Boil water in a pot. Pour in the sweet corn kernels and boil for a while. Add cream-style sweet corn and bring to the boil again. Pour in the minced chicken from step 1 while stirring continuously to break it into tiny bits. Do not cover the lid.

**4.** Bring to the boil again. Put in fish tripe. Season with a pinch of salt. Cook till the fish tripe turn slightly translucent. Turn to medium heat and stir in egg white. Pour in the thickening glaze while stirring continuously. Sprinkle with shredded Jinhua ham on top. Serve. *pictures 1~4*

### GRANDPA'S TIPS

- *Minced chicken meat is quite sticky. That's why I add water to it and stir it into a paste so that it won't stick together in big lumps when added to the soup.*

- *Chinese thick soup is always thickened with water chestnut starch because it holds its consistency better than other starches. The soup won't turn watery even after it's been over-stirred.*

- *Sand-puffed fish tripe is less greasy than deep-fried ones. It is easy to handle and it gives a crunchy mouthfeel without being sticky after cooked.*

- *When you make any glaze with egg white, do not cook it over high heat. Otherwise, the ribbons of egg white will overcook and turn tough.*

- *Here's a trick to soak dried scallops. Peel the tough tendon off the dried scallops (it looks lighter in colour). Put dried scallops into a dish. Pour water to just cover them. Then leave them until no water is visible on the dish. Steam for 30 to 35 minutes. Turn off the heat and leave them in the steamer with the lid covered for 10 to 20 more minutes. Then just take what you need for the recipe and crumble them into shreds before use. Quality dried scallops should be in whole with as few cracks as possible. Bigger ones tend to be more flavourful and softer in texture.*

## INGREDIENTS:

1 head mustard greens
1/3 head salted mustard greens
300 g pork ribs
1 frozen pork tripe
2 tbsp dried soybeans
10 g white peppercorns

## METHOD:

**1.** Tear the leaves off the mustard greens one by one. Rinse well and cut into chunks. Set aside. Soak the salted mustard greens in water for 20 minutes. Drain.

**2.** Chop the pork ribs into pieces. Blanch in boiling water. Drain and set aside. Thaw the frozen pork tripe. Turn it inside out. Rub coarse salt on it to remove the fat and dirt. Rinse well. Blanch in boiling water. Drain and cut into chunks.

**3.** Crack the white peppercorns with mortar and pestle. Soak soybeans in water till slightly softened.

**4.** In a pot, add pork ribs, pork tripe, salted mustard greens, fresh mustard greens, soybeans and water. Bring to the boil. Add white peppercorns. Turn to medium heat and boil for 1.5 hours.

# PORK TRIPE SOUP WITH PEPPERCORNS AND MUSTARD GREENS

*Refer to p.125 for steps.*

## GRANDPA'S TIPS

◆ *Quality fresh mustard greens should have a rounded base and thick leaves. When you tear a leaf off it, you should smell a peppery fragrance.*

◆ *The white peppercorns just need to be cracked. You don't need to grind them finely.*

◆ *Taste the salted mustard greens after soaking it in water. Sometimes they are too salty to be eaten. To make sure the soup won't be too salty, add mustard greens in moderation.*

◆ *Blanching the pork ribs not only removes the blood, but also seals in the flavours by shrinking the meat on the surface.* **pictures 1~3**

◆ *Soybeans give the soup soy flavour and make it more nutritious.*

# PORK AND CHICKEN SOUP WITH CONCH, SHA SHEN, YU ZHU AND DRIED LONGANS

## INGREDIENTS:

5 small frozen conches
225 g lean pork
1/2 chicken
dried tangerine peel
Sha Shen
Yu Zhu
Huai Shan          *
dried longans
Goji berries
38 g candied dates

*Tell the herbal store that you need ingredients for 4 servings.*

## METHOD:

**1.** Rinse the chicken and pork. Blanch in boiling water separately. Drain. *pictures 1~2*

**2.** Remove the opercula of the conches. Cut them open and remove the innards. Cut open the siphons and clean any dirt inside. Rinse again and blanch in boiling water. Drain. *pictures 3~8*

**3.** Soak dried tangerine peel in water till soft. Scrape off the pith.

**4.** Soak Sha Shen, Yu Zhu and Huai Shan in water briefly. Rinse well. Rinse the Goji berries.

**5.** Put dried tangerine peel, pork, conches and chicken into a soup pot. Add cold water. Bring to the boil over high heat. Add Sha Shen, Yu Zhu and Huai Shan. Boil over medium heat for 1.5 hours. Add candied dates and dried longans. Turn to low heat and simmer for 1 hour. Put in Goji berries. Boil for 30 minutes. Serve.

### GRANDPA'S TIPS

◆ *Some people may cut off the conches' siphons before use. But I always keep them when making soup. They taste crunchy and delicious.*

◆ *Whenever I make soup with pork, chicken or conches, I always blanch them in boiling to remove any dirt, blood and fat. The soup would end up clearer that way.*

◆ *The order of adding ingredients to the soup depends on how ready an ingredient is for its taste to infuse in the soup. I put in those dried ingredients and meat first as it takes time for them to render their flavours. Ingredients that lend their taste quickly to the soup, such as dried longans and Goji berries, can go in at last.*

## INGREDIENTS:

1 small white cabbage

4 dried scallops

38 g large dried shrimps
(rinsed; soaked in water briefly)

1 dried octopus (rinsed; soaked in water till soft;
skin peeled off; cut into strips)

8 dried oysters (rinsed; soaked in water briefly)

2 cubes Jinhua ham

4 dried shiitake mushrooms
(soaked in water till soft; sliced)

1 canned abalone (sliced)

6 medium prawns
(with legs, antennae and rostrums trimmed off)

## METHOD:

1. Cut off the base of the cabbage. Peel of a few wilting leaves on the surface. Cut out the core with a small knife. *pictures 1~5*

2. Boil water in a pot. Put in dried scallops, dried shrimps, dried octopus, dried oysters, Jinhua ham, shiitake mushrooms and abalone. Boil for 30 to 40 minutes.

3. Put in the medium prawns and do not cover the lid. Wait till the prawns are half-cooked. Put the cabbage in with the cut facing down. Cover the lid and cook for 20 minutes (for a small cabbage). Serve. *pictures 6~7*

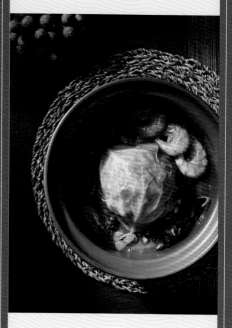

# BRAISED CABBAGE IN GOURMET SEAFOOD SOUP

*\* Refer to p.130 for steps.*

### GRANDPA'S TIPS

◆ The core of the cabbage has to be remove before use. It's because the core is too tender and may break down after prolonged cooking. The soup may turn cloudy and less appetizing that way.

◆ I put the cabbage into the soup with the cut side down because it's easier for the cabbage to pick up the flavours from the soup that way. Both the cabbage and the soup are the heroes of this recipe and they taste equally great.

◆ Make sure you trim off any black mouldy skin on the Jinhua ham. They tend to taste like rancid oil. *pictures 8~12*

# GAROUPA BONE SOUP WITH BEAN SPROUTS, TOFU AND TOMATO

## INGREDIENTS:

600 g garoupa bones
300 g soybean sprouts
1 tomato
1 cube firm tofu (diced)
2 slices ginger
salt

## METHOD:

1. Boil water in a pot. Put in ginger and garoupa bones. Boil over medium-high heat until the soup turns milky. Put in soybean sprouts and boil for 30 minutes.

2. Add tomato (it usually floats and it helps to press it to the bottom of the pot with a ladle or chopsticks). Add tofu and cook for 20 minutes. Season with salt.

### GRANDPA'S TIPS

- *When you make soup with marine fish bones, you don't need to fry it in oil before use. Yet, when freshwater fish bones are used, always fry them in oil until golden because they may have a muddy taste.*

- *If you want to make two courses with one fish, you can fillet the garoupa and stir-fry the fillet with broccoli and yellow Chinese chives. The fish bones can be used to make this soup.*

- *I put a chopstick on the rim of the pot before putting the lid on so that the pot is not fully closed. That would prevent the soup from boiling over. see p.133*

## INGREDIENTS:

6 hard-boiled eggs
300 g gingkoes (shelled)
2 sheets dried beancurd skin
2 whisked eggs
rock sugar to taste

## METHOD:

**1.** Boil water in a pot. Put in gingkoes and cook for 20 minutes. You can easily remove the skin. *picture 1*

**2.** Boil water in a pot. Add rock sugar and cook till it dissolves. Taste and add more sugar if needed. Put in skinned gingkoes and boil for 10 minutes. Add beancurd skin till it breaks down completely. Put in the hard-boiled eggs. Stir in whisked egg while cooking. Turn off the heat. Serve. *pictures 2~5*

# BEANCURD SKIN SWEET SOUP WITH GINGKOES AND EGGS

# WHITE FUNGUS SWEET SOUP WITH PAPAYA AND BANANA

*\* Refer to p.137 for steps.*

### GRANDPA'S TIPS

◆ This retro sweet soup facilitates bowel movements and smoothens the skin.

◆ I prefer half-ripened bananas for this recipe.

◆ When I make soup or sweet soup with papaya, I always choose those "male" papaya with a slender shape and fewer seeds.

◆ For this sweet soup, put in papaya and banana in the opposite order of their ripeness. That is, if the papaya is less ripe than banana, papaya should go in first, and vice versa.

## INGREDIENTS:

1 ripe papaya
rock sugar
2 bananas
(peeled; cut into chunks) *picture 1*
1 head fresh white fungus (soaked till soft ; remove the tough root; tear into florets)
19 g Goji berries

## METHOD:

**1.** Peel, de-seed and cut papaya into chunks.

**2.** Boil some water and add rock sugar. Cook until sugar dissolves. Taste it and add more water or sugar if needed. Put in the papaya and cook for 3 minutes. Add banana and cook till it start breaking down. Add Goji berries and white fungus. Bring to the boil. Serve. *pictures 2~5*

# Master the *Art of Cooking*

掌握烹調的藝術

Towngas KIDchen
煤氣親子烹飪班

Demonstration Room
烹飪示範班

Practical Room
烹飪實習室

Hassle-free Enjoyment
盡情享樂體驗

Well-equipped Facilities
設備完善

Professional Tutors
師資優良

銅鑼灣波斯富街99號利舞臺9樓 9/F, Lee Theatre, 99 Percival Street, Causeway Bay 電話 Tel : 2576 1535

   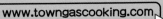 www.towngascooking.com

# 明火煮意系列
## *Flame Cooking Series*

明火專用
For **Flame** Cooking

## 魔法烘烤板
## Multi-FUNctional Magic Pan

**樂** Joy 3合1模板可製作多款窩夫和鬆餅
3-in-1 mould plan for creating waffles and pancakes

**易** Easy 容易更換模板，方便清理
Cleaning and changing moulds are easy

## 韓式燒烤架 Korean Grilled Pan

**易** Easy 不黐底貼心設計
Non-stick thoughful design

**快** Fast 傳熱快令食物更惹味
Faster heat distribution enhances food tasty

## 不銹鋼多用途蒸鍋 Stainless Steel Multi-Steamer (26厘米cm)

**省** Save

**營** Health

全線煤氣客戶中心，煤氣烹飪中心、名氣廊及www.towngasshop.com有售
Available at all Towngas Customer Centres, Towngas Cooking Centre, Towngas Avenue and www.towngasshop.com

## 煤氣
## Towngas

www.towngasshop.com

f Towngas Green Living

# CROWN® 皇冠牌
## —since 1979—
# CGS

## 廚具系列

**CE-766易拆式抽油煙機**

**CE-723電熱除油抽油煙機**

**CE-937煙導掛牆式抽油煙機**

## CGS銷售點地址：

西灣河成安街2號地下
G/F, 2 Shing On Street, Sai Wan Ho, HK

香港灣仔灣仔道 83 號地下 1 號 B 舖
Shop 1B, G/F, 83 Wan Chai Road, Wan Chai, HK

柴灣杏花村杏花新城東翼商場 2 樓 202A 舖
Shop 202A, 2/F, Paradise Mall, East Wing, Heng Fa Chuen, Chai Wan, HK

長洲大新海傍路107號B地下
G/F, 107B Tai San Praya Road, Cheung Chau, HK